Flower in the Crannied Wall

Flower in the crannied wall,

I pluck you out of the crannies,

I hold you here, root and all, in my hand,

Little flower, but if I could understand

What you are, root and all, and all in all,

I should know what God and man is.

Alfred Lord Tennyson

Let's Take A Walk

Discover the habitats, adaptations, uses and folklore of some common wild plants of Eastern United States

by:
Win Carter

and illustrations by:
Pamela Steele Carter

Let's Take a Walk.
Copyright © 1991
by Gale Winston Carter

Library of Congress Catalog Card Number
90-093571

ISBN # 0-9628153-0-6

PRINTED BY:

to my dear wife, Pam

who has been an important collaborator

in the preparation of this book.

Acknowledgements

A number of individuals have helped to make this book a reality, and I am indebted to them for their help and influence.

Leonard Bradley, Audubon staff botanist for many years. As a colleague, Leonard helped me to get in touch with the deep satisfaction to be found in botanical field work.

Robert Paier, Editor of the Connecticut Environment Magazine for his support over the years. Much of the information for this book as been drawn from articles that I have written for "Connecticut Environment."

Dr. William A. Niering, Professor of Botany, Connecticut College, for reading the galley proofs and for his many helpful suggestions concerning my manuscript.

Mr. & Mrs. Donald Swan of the Connecticut Botanical Society for reading parts of my manuscript and furnishing up-dated information on scientific names of plants.

Patricia Dalton Haragan, botanist for the University of Tennessee, and members of the Audubon staff in Greenwich for their contributions.

Foreword

In Zen Buddhism, probably the highest compliment you can pay a person is to say that he "leaves no traces". The idea is that such a person has gone beyond small-minded exaltation of the ego, to personal self, and now encounters nature cleanly, clearly, on its own true and pure terms. Such a person walks through life, through nature, quietly, without pretense, with absolute integrity. I believe that Win Carter walks through nature leaving no traces. And, I believe the reader of this book of walks is in for a rare and wonderful treat - to see nature the way Win Carter does, in great detail, with utter openness, with perfect honesty. Learning about the plants that grow so profusely and magically around us, their history, their wonderful uses, is just the beginning of what the reader will derive from this book. Really, for me at least, what Win Carter is writing about is how to walk through nature leaving no traces.

My association with Win Carter goes back six years. It was at that time that I became editor of a magazine called *Connecticut Environment,* published by the Connecticut Department of Environmental Protection. As the new editor, I had lots of new ideas, new ways of jacking up the pzazz of the magazine. One of the columns in the magazine, which had run since 1976, was "The Trailside Botanizer". Each month, "The Trailside Botanizer" would describe in meticulous detail one single plant or shrub or wildflower, talking about its history and uses. "We're going after big environmental issues here," I said. "Do we really want a monthly column on wildflowers?" The response was a shrug. "Maybe you should read it for a while. People really seem to like it". So, for a few months, I just watched "The Trailside Botanizer". The articles had a quiet charm; there was something about them that gently, delicately, drew you in. There was a sense of being in a different world with each article, a world of sanity and order and harmony. After a few months, I also was one of those many people who really liked "The Trailside Botanizer".

As more time went by, I found that no matter what other kind of material I ran in the magazine, and some of it has been rather good, "The Trailside Botanizer" has continued to be the single most popular column in the magazine. How do I know this? Simple. Readers write in and tell me.

I am proud to know Win Carter. I am honored that he asked me to write these few words of introduction. Most of all, I am grateful for his teaching me to see a part of the world the way it really is - cleanly, clearly, honestly, without affectation. For anyone who has not yet read Win Carter's descriptions of nature, you are in for a very special experience. To his many ardent fans who have awaited the publication of this book, your patience has been rewarded. And to everyone about to take a walk with Win Carter, prepare yourself for the remarkable experience of walking through nature with a man who leaves no traces.

Robert Paier, Editor
Connecticut Environment Magazine

Introduction

This book is written for the novice who may never have purchased a field guide as well as for the more knowledgeable natural history buff. It is designed to help you take a closer look at plants found in familiar habitats.

My book will take you, the reader, on a personally guided tour to a number of different kinds of environments. Many of these habitats are found within a short distance from where you live. Perhaps you have visited these places many times and missed the great beauty of the usual and commonplace in nature to be found there.

The complexity and beauty of a flower is a stimulus to learning. You are encouraged to learn the name of the plants that you find. The common name may have an interesting story to tell, while the scientific name is often descriptive.

Knowing the name of a plant is the first step in getting better acquainted, just as with a new friend, and it brings us closer to understanding a particular flower, its value and role in the environment. Now we are better able to delve into the folklore, adaptations and possible uses of each new plant we discover.

To become more familiar with a new-found flower, we will want to know some of its special characteristics, its habitat and the associated plants growing nearby. This information is found throughout the book.

You can use this book in a number of ways. You can start very close to home, in fact right on your own lawn, or with less familiar surroundings. You may want to take a seasonal approach, as some walks are best taken at a particular time of year.

These walks can be enjoyed in private or in a group, using this book as a guide. You are encouraged to bring such items as notepaper, sketch pad and camera to preserve a record of what you see.

Common plants to be found on each trip are highlighted. When you return from your walk, you may wish to refer to some of the line drawings or look up some specific data in the text which covers 75 common plants in detail.

The information on asters, goldenrods and violets is intended especially for those who are more technically minded.

My purpose in taking you on these walks is to deepen your pleasure and understanding of the so-called "commonplace" in nature. Knowledge and understanding will help us develop a protective attitude toward our constantly threatened natural resources. If this book contributes in any way to such an attitude on your part, it will have more than fulfilled its purpose.

Contents

1

A Walk Along a Stream in Early Spring

BY THE LATTER part of March, the time of the vernal equinox, certain forces in the natural world are put into motion that cannot be reversed. The days are becoming longer and the nights shorter. There is a feeling in the air, an indefinable sensing that the worst of winter is behind us. It is a day in spring, but not necessarily what we may call a spring day.

March 21st is officially the first day of spring in all parts of the country, yet we find conditions vastly different from state to state, from town to town, and even in the somewhat confined area of a woodland.

There comes a time, usually towards the end of March, when the longer days bring with them melting snow and the rise of sap in the trees. Nostalgia takes me back to former days when I was a high school student at an academy in northern New Hampshire. At this time of year many students skipped school to visit a sugar house in the woods. There was a great deal of interest in watching the sap from the sugar maple being boiled down into maple syrup. Because there was so much interest in this event, a day was set aside at school so that everyone who wanted to watch this process would have a chance to see it. One of the highlights of this trip was the opportunity to have sugar-on-snow. Hot syrup was poured on fleecy white snow. The tacky sugar was then wound up onto a wooden paddle. This produced a huge "sugar-daddy." There was nothing quite like it.

Tree sap, which has been stored in the roots during the winter, must move up the tree to provide the energy for the eventual bursting of winter buds. In March, on a bright sunny day after a cold night, the sun warms the bark and wood sufficiently to start the upward movement of sap in the sugar maple, which is one of the earlier trees to respond to the increase in the amount of daylight.

The early colonists learned from the American Indian how to harvest maple syrup and maple sugar from the sugar maple. The Indians made gashes in the trunk of a sugar maple and collected the clear sap. The colonists devised some new techniques. They drilled holes in the tree, inserted the spouts of hollow elderberry twigs and then boiled down the sap in kettles.

With special footwear to allow for muddy walking, I like to look for early signs of spring along a favorite stream at this time of year. This trail is less than a mile in length, but passes through several habitats with different kinds of vegetation. Our walk begins in a wooded area.

Here, the water runs high in the stream and collects in little backwaters, producing a soft, continuously recycled mud, rich in organic materials. Each year, along this stretch, I look forward to finding the first signs of growth in the skunk cabbage and false hellebore. Even as early as February, there may be evidence of early growth and blossoming in the skunk cabbage. Both plants push their way through the frozen mud along the stream and in muddy depressions nearby. Skunk cabbage requires more water, so it often grows right in water, while the false hellebore seems to prefer wet soil somewhat away from the stream rather than areas that are constantly flooded.

As we walk into the woods, the next evidence of spring to catch our eye is little dabs of yellow sprinkled profusely throughout the moist woodland and along both sides of the stream.

Few plants give more of a feeling of spring than the spicebush in bloom. During the long winter months, the buds with the embryo leaf and blossom within wait patiently. They were formed long before cold weather had set in, and wait for that moment in early spring when the buds are triggered to open.

Overhead, an American elm is beginning to flower, producing its tassel-like blossom. We usually think of elms as street trees, but their natural habitat is along streams and in wetlands. The slippery elm has a similar natural habitat. Its flowers resemble those of American elm, but the clusters are somewhat tighter.

A few days earlier along other waterways the silver maples have started their rush toward spring by sending out miniature yellowish-green flowers. They are the earliest of the maples to blossom, but the red swamp maple is not far behind. Red maple begins to produce its profuse red flowers about the same time as the elms.

The trail along the stream twists and turns until it comes to a clearing. Last season's grass and summer flowers have changed their color and appearance. Protruding from the snow that remains are stalks of grass now tan, broken and bent. The flowers of the past year are but a memory. The only evidence that they once brightened this area is the brown remains of seed pods, long since emptied of their contents. Nearby, the stream banks are thick with alders and willows and there is a stirring of life as their catkins begin to emerge.

In the natural world, what appears to be lifeless suddenly seems to be reborn. Those who anxiously watch for subtle signs of spring in the plant world will find them in the pussy willow. This species has larger pussies, or undeveloped flower clusters, than other species of willow. Even in February, some of these plants are beginning to show a bursting of their single scale buds and the emergence of the fur-like pussy, which appears before the leaves. Individual plants may be found in blossom well into May.

The catkins of the willow show some interesting examples of adaptations to weather conditions. The single scale of the bud protects the developing flowers from evaporation. When the scale falls off, the many hairs that surround the florets are further protected from too rapid evaporation and from the drying winds.

The smooth alders, which grow as a shrub or a small tree, often go unnoticed during the summer months because there is little about them that is colorful. In the winter, however, the persistent cones, or strobiles, become more noticeable and in the early spring, the male and female catkins begin to expand and flower. This occurs before the leaves emerge. The catkins, which contain many small flowers, or florets, were formed the previous season and wait for favorable conditions before blossoming. The flowers are protected by dark, overlapping purplish-red scales in the winter. The staminate clusters of catkins expand several times their original length when they blossom.

The flowers of elm, willow and alder all blossom before their leaves appear. They are wind pollinated and lack the colorful petals of many flowers. These are all important adaptations to allow the escape of pollen without the obstacles of petal and leaf.

As we come to the end of our walk. we realize that March doesn't have many signs of spring, but what it does have is special. All of the early signs of emerging life that we have seen create the stage that fuels our anticipation of a much more active explosion of new green leaves and blossoming to come.

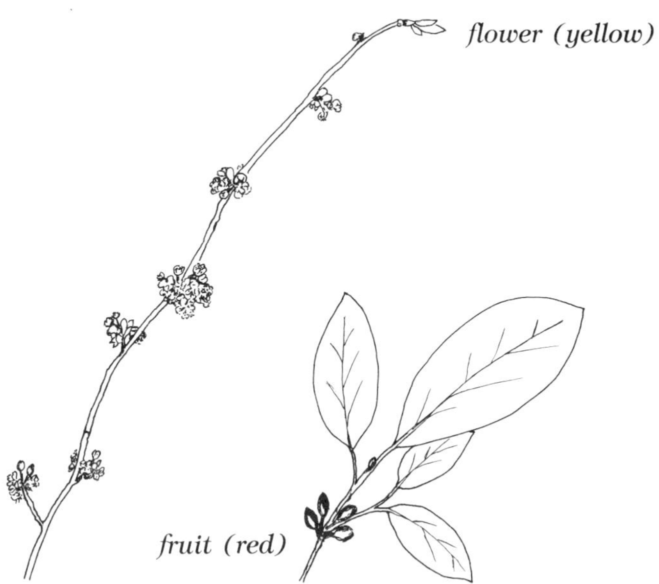

Spicebush

(Lindera benzoin)
Laurel Family

Size: 3-12′

Spicebush, along with the pussy willow, is an early harbinger of spring. It is usually thought of as a wetland plant that grows as an understory shrub in moist woodlands and along streams.

The blossoming period of the spicebush is often as early as March, but generally lasts into April. The tiny flowers appear before the leaves. They are unisexual, which means that the male and female blossoms are located on separate plants. The flowers lack petals. What we see that resembles petals are really yellow sepals. The flowers emerge from globular flower buds which are different from the smaller pointed leaf buds. The red fruit ripens between July and October.

All parts of the plant are fragrant if crushed or scratched. Its odor is often described as lemony. The smell resembles that of gum benzoin obtained from a native tree of Java and Sumatra. This may explain the origin of the species name "benzoin."

The twigs and leaves of spicebush can be used for making a tea, while the berries were once used in powdered form when allspice seasoning was not available.

The fruit is the preferred food of the wood thrush and veery, but is eaten sparingly by quite a few other songbirds. Spicebush is also the host plant of the spicebush swallowtail butterfly.

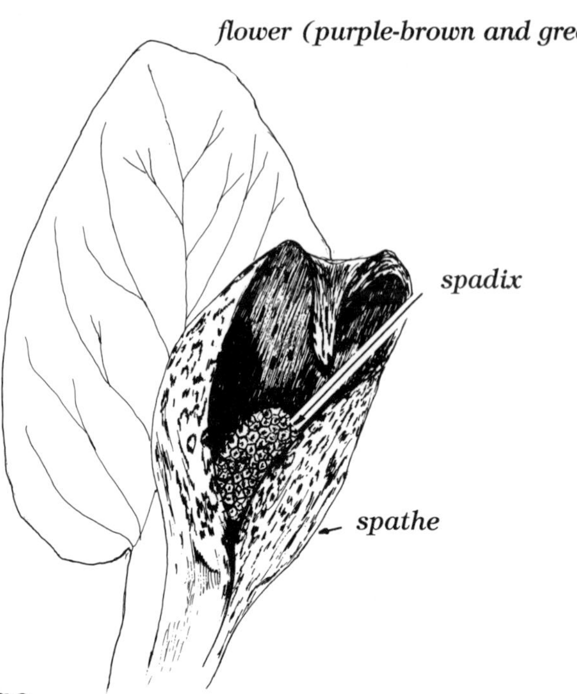

Skunk Cabbage

(Symplocarpus foetidus)
Arum family

Size: 1-2′

The flowers of skunk cabbage are usually the first of the early spring wildflowers to bloom, sometimes as early as February. The first signs of growth appear as red, green or brown horns sticking up through frozen muck. These are commonly found along stream banks and the edge of swamps. The strange looking horns are the flowers of skunk cabbage formed the previous fall. As the winter begins to wane, the skunk cabbage begins to generate heat as its growth processes quicken. This enables it to push its way up through the ice and snow. The heat generation also favors the spread of odor and attracts insects.

Individual flowers of skunk cabbage are tiny and are crowded onto a ball-like structure (the spadix) which is enclosed in a greenish to reddish brown, slipper-like hood (the spathe). Each flower consists of four sepals, four stamens and one pistil. Pollination is believed to be carried out by carrion flies and small gnats. The fruit is the enlarged spadix that becomes spongy and filled with seeds.

The skunk cabbage gets its name from the resemblance of its young leaves, which emerge after pollination, to those of cabbage and the odor of its juice to the offensive scent produced by skunks.

Skunk cabbage is an edible vegetable if properly prepared to remove its odor and irritating calcium oxalate crystals.

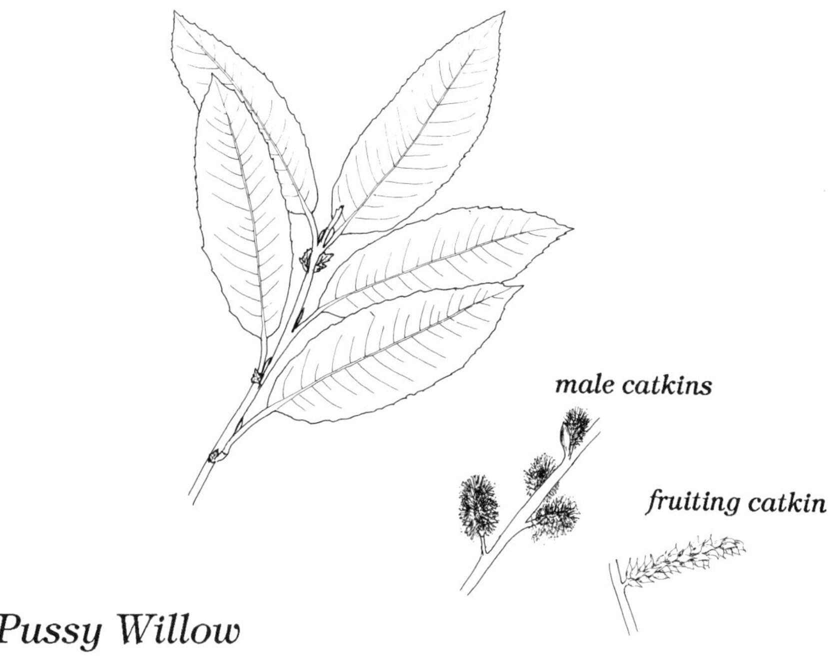

Pussy Willow

(Salix discolor)
Willow family

Size: to 30′

In early spring, it is the developing flower of the pussy willow that attracts our attention. Pussy willows usually grow as a shrub or small tree. They prefer a wet habitat such as the edge of a swamp or riverbank. The male and female flower clusters, which are called catkins are borne on separate plants.

As early as March, the shiny reddish-purple, single-scaled buds begin to open revealing the well-known fur-like pussies. These appear before the leaves come out. When the male catkins mature, they become covered with golden pollen which bees use to feed their young. The female catkin is softer, longer and light green in color. Both male and female catkins consist of many tiny flowers or florets. A nectar-bearing gland is attached to both sexes to attract bees. The pussy willow is pollinated by both wind and insects.

The fruit can be observed from April to May and is a small urn-shaped capsule. Each capsule contains many seeds, each equipped with a tuft of silky hair which adapts them for wind dispersal. These capsules appear as drooping catkins.

The hairy seeds are used by the yellow warbler for nest building, while ruffed grouse relish the winter buds.

The bark of willow trees contains salicylic acid, a chemical used in the making of aspirin.

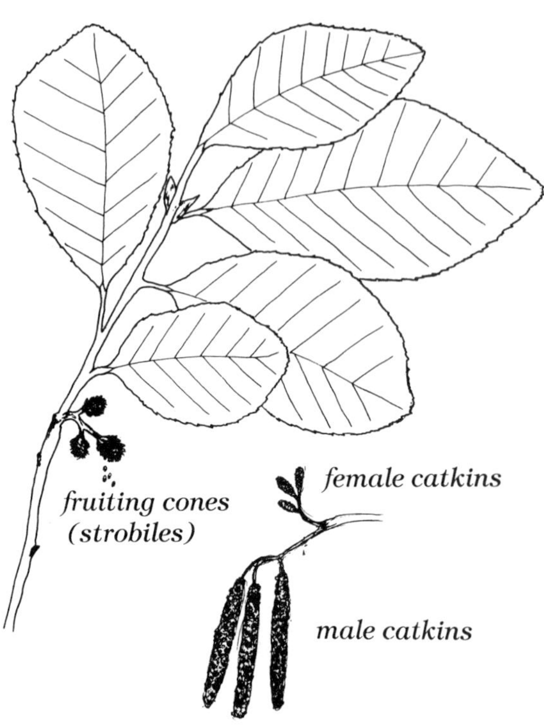

Smooth Alder

(Alnus serrulata)
Birch family

Size: 6-12′

This early blooming shrub, or occasionally small tree, is commonly found growing along streams and in wet meadows. It is closely related to the birches. Both alders and birches have similar catkins.

During the flowering period, which is from March to May, the pendant staminate catkins appear golden and purple as they shed their load of pollen. The pollen matures slowly and is carried by the wind to the female catkins which are on the same plant. The female clusters are very small and cone-like and expand only slightly as they blossom. They are usually found directly above the male cluster of catkins. When blossoming occurs, very small reddish hairs project outward in all directions from each tiny flower.

The woody fruits that form open in the autumn to shed their tiny seeds which are dispersed by the wind. The cones, or strobiles, remain on the tree and do not disintegrate. Birch trees have the same type of cones, but these disintegrate and do not persist on the trees.

Later in the season, red-winged blackbirds and goldfinches will nest in the upper branches. Chickadees and goldfinches will eat the seeds of the fruit, while grouse will feed on the reddish, stalked buds. Rabbits and deer occasionally browse on the twigs.

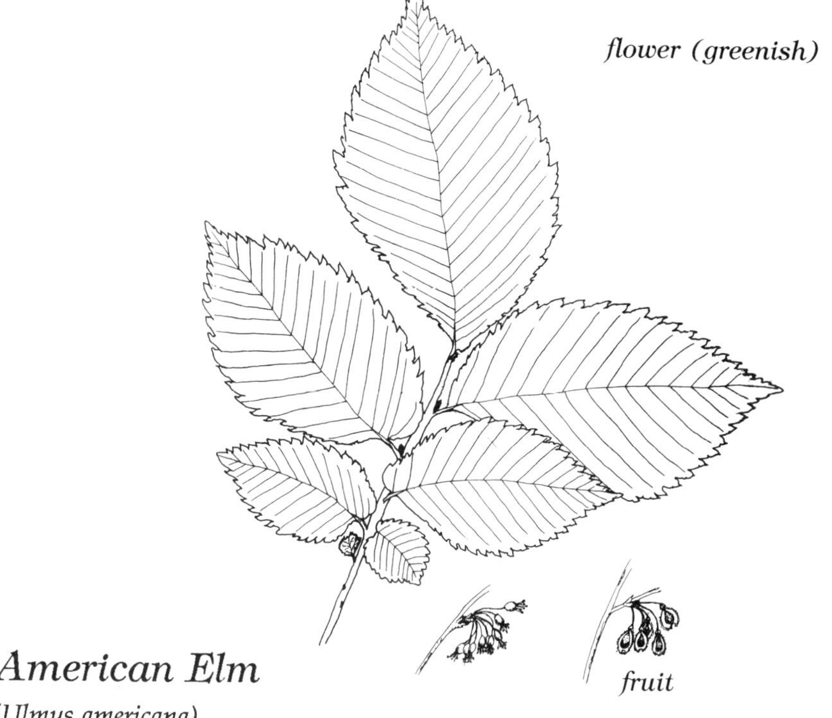

American Elm

(Ulmus americana)
Elm family

Size: 80-100′

The American elm is the best known of our elms. In the past, it has played an important role in making our New England towns picturesque. It was a common sight on village greens and as a street tree before the appearance of the Dutch Elm disease. Throughout our history it has served as a ceremonial tree.

The species takes several forms, but the most common and familiar is the vase-shaped, or fountain-like, appearance. Although usually thought of as a street tree, American elm is more at home in wetland areas along streams and rivers.

American elm has deep-furrowed bark separated by flat-topped ridges. Its leaves are slightly lop-sided and sometimes sandpapery. The usually smooth dunce-cap winter buds have reddish brown scales rimmed with dark edges. The twigs often have a zigzag appearance.

The flowers bloom early in February before the leaves, and are in slender dangling clusters. Each flower has both stamens and pistils. The small, wafer-like seeds form soon after flowering. This is about the time the leaves appear.

These seeds are eaten by gamebirds such as the ruffed grouse, and rodents, like mice and squirrels. Deer and rabbits browse on the buds and twigs.

Chapter 1

Plants Along A Stream

American Elm - *Ulmus americana*
False Hellebore - *Veratrum viride*
Pussy Willow - *Salix discolor*
Red Maple - *Acer rubrum*
Silver Maple - *Acer saccharinum*
Skunk Cabbage - *Symplocarpus foetidus*
Slippery Elm - *Ulmus rubra*
Smooth Alder - *Alnus serrulata*
Spicebush - *Lindera benzoin*
Sugar Maple - *Acer saccharum*

2

A Walk in the Spring Woods

SPRING STARTS OFF with only a few signs of greenery and then seems to gain momentum, like a moving picture film when the frames have been accelerated. As the snow melts from the woodland, there is a smell of fresh earth in the air. Everything seems to be in sharper focus. There is a clean, scrubbed look to the woods. There is a feeling of anticipation as to what the first discovery will be.

Between the early thawing of snow and the appearance of leaves on the trees, a number of early perennial spring flowers must send up leaf and blossom quickly. The melting snow, the increased amount of light, the higher temperature and moisture content in the air all combine to stimulate plants into rapid growth. Taking advantage of the full sunlight before the overhead canopy develops, they must, in a brief period, manufacture and store enough food in the underground storage organs for quick growth and flowering the following spring. This adaptation is thought to be one of the legacies of the Ice Age. There had to be a quick way of producing flowers and seeds during the very brief summer. This was and is the solution.

These delicate plants are called spring ephemerals. After they produce flowers, their foliage dies back and no evidence of plant growth is seen until the following spring. They flourish in the leaf mold which holds more moisture than is found on bare ground in open sunlight.

Trout lily is the first of the ephemerals to flower. It is followed by such plants as spring beauty, rue anemone, Dutchman's breeches, squirrel corn and cut-leaved toothwort. Ephemerals are often found growing under oaks and hickories which are among the last of the trees in the spring to leaf out.

Some of the common spring flowers, such as bloodroot, wood anemone, hepatica, violets, may apple, wild ginger, Jack-in-the-pulpit and trillium are not considered ephemerals. They are shade plants with foliage that has the ability to continue manufacturing food in reduced light after the canopy closes overhead.

Spring wildflowers grow in many different habitats in the woods, but are most common on lower slopes in moist hardwoods of maple and beech. Heavy leaf litter in many oak forests is not favorable for wildflowers.

Many wildflower species are in bloom at this time of year. The flowering period extends over three months to include both late and early blooming species. Mid-May is a good time to see some of the early flowers that are still in bloom and to catch some of the flowers that are just beginning to blossom. Frequent hour-long trips to a favorite spot will help you to see growth changes which will help to give a better understanding of each plant.

To become familiar with a great variety of species, a larger area of land should be selected. Places like nature centers, sanctuaries, state parks and state forests are ideal because of the possibility of extending the time frame for walking and observing.

One of my favorite sports for a short walk is our local nature center. Here, I only have to walk a short distance before I begin to see spring flowers. By Mid-May, the maples and birches have unfolded their tender new leaves, which are showing pink and varying shades of green. The shadbush is in bloom and the highbush blueberry is dotted with many tiny pinkish, bell-like flowers. The cinnamon ferns have lost their fiddlehead look and are nearly unfolded.

Low on the ground, in a clearing near the brook, arising through a stand of haircap moss is the tiny dwarf ginseng. Its white flowers are arranged in a tight, rounded cluster called an umbel. It looks much like true ginseng, long valued by the Chinese for medicine. Dwarf ginseng is rather common and is much smaller. Its toothed, compound leaflets are arranged in a whorl around the stem. The flowers appear from April to June. These are followed by the development of yellowish berries arranged in clusters.

Nearby there are many small orchid-like flowers, fringed polygala, often called "gaywings" or "flowering wintergreen." This flower is not an orchid. It belongs to the milkwort family. The usual habitat is in rich or rocky woodland, although I have found it growing in an alkaline bog, and it is sometimes found in dry sites.

Along the brook, there are a number of sensitive ferns, so called because they are one of the first ferns to be hit by the frost in the fall. Scattered near them are some of their last year's spore cases which grow on separate stalks. At the top of each stalk are the brown, bead-like spore cases which have long since discharged their contents.

As we continue walking along the brook, we can see numerous specimens of yellow rocket, or winter cress, an early member of the mustard family, and two kinds of anemone. Both anemones belong to the buttercup family and each has a rather long blooming period.

Rue anemone is one of the spring ephemerals and its scientific name. *Anemonella thalictroides* means "little anemone resembling the meadow rue," which refers to the leaves. The rue anemone is not a true anemone because it belongs to a different genus, but it does resemble the wood anemone. The Ancients had the belief that these flowers would not open until they were stirred by the gentle spring winds. The early Saxons called the wood anemone "flaw flowers," flaw meaning "gust." They wave even in the slightest gust of air.

There are physical and behavioral differences in these two kinds of anemones. A behavioral difference between the two flowers is that the flower of the wood anemone closes when the day is overcast, while those of the rue anemone remain open.

Physically, the stem and leaf stalks of rue anemone resemble fine black wire. The leaves are rounded with three lobes. They are arranged in whorls of three. The small flower lacks petals, but has five to ten white sepals and many stamens and pistils. Look carefully at the leaves of both rue and wood anemone to notice how they differ in this and perhaps other ways.

Another member of the buttercup family grows in the water near the edge of the brook. This is the marsh marigold. Swamps and marsh areas are other favorite habitats for this showy flower.

The buttercups are well represented among the spring flowers. The family is a very ancient one and among all the flowering plants, it appears to have changed least from its early ancestors.

A short distance from the brook, growing under some trees, are large patches of trout lily. Only the leaves remain now as they are one of the earlier flowers to bloom, being a spring ephemeral. Soon, even the leaves will disappear. Trout lily and Dutchman's breeches, another ephemeral, are good examples of the great adaptability of plants to changing temperatures. They have a great deal of sugar in their leaves, which makes it possible for them to endure hard freezes for more than twenty-four hours before the frozen sap eventually kills the cells.

On the way back, we move to another trail that is rich, moist woodland and less open. The first discovery is wild geranium, or cranesbill. It is one of the lovelier flowers to blossom during the spring and early summer. Here, there are only scattered specimens of those rose-pink to magenta flowers. They make a more dramatic showing when they appear in clusters.

Jack-in-the-pulpit seems to prefer the same type of habitat as the wild geranium. This is true of this site we are exploring, where we find several of these handsome plants. Jack-in-the-pulpit is one of the more familiar of the spring wildflowers. It usually blooms from the middle of May to early June, and is easily identified because of its unusual flower.

Another flower that grows in the moist-rich woodland, is the red trillium. The showy red petals will attract one's eye before we notice the strong odor as we kneel to admire it.

In the more open woods, on our return trip, we are able to locate a few pink lady's slippers growing under some pine trees. At times, on other sites, these beautiful orchids appear in huge numbers. Occasionally, it is possible to locate one that is white. Pink lady's slippers are becoming more common in areas that are returning to woodland. They reproduce most vigorously in light shade and some moisture, but are found in all conditions except where there are some extremes of shade and sun.

There may be a variety of habitats present in your own community where you can discover some of the flowers I have mentioned. You can extend your exploration even more by visiting limestone-rich sites and areas that have acid bog conditions. The joy of discovery enhances the pleasure to be found in a spring walk.

Trout Lily

(Erythronium americanum)
Lily family

Size: 4-10″

The flowers of this small woodland lily may appear along mossy stream banks or the edge of bogs as early as March, but blossoms continue to form into May.

Its mottled, green, lance-shaped leaves give rise to common names such as trout lily and fawn lily. The spots on the leaves were thought to resemble those of a brook trout or a young fawn. The naturalist, John Burroughs, thought that the two leaves of trout lily stood up like fawn's ears.

The reflexed yellow flower of trout lily consists of three sepals and three petals, which are nearly identical. Six anthers project outward from the flower to resemble what some think look like an adder's tongue - a reason for another of its common names.

The plant is pollinated by queen bumblebees, flies, solitary bees and early butterflies. It may also be propagated by small bulbs that are outgrowths of its underground, bulb-like corm.

Trout lilies with only one leaf will not blossom. Two leaves must be present in order for it to flower and it takes several years for this to happen.

The Indians boiled and ate the small bulbs. The young leaves were used for greens, soups and stews and for making a tea for relieving hiccoughs and vomiting.

Pink Lady's Slipper

(Cypripedium acaule)
Orchid family

Size: 6-15″

The pink lady's slipper, or moccasin flower, is an example of one of our common native orchids. This orchid has a special beauty which many plant lovers look forward to seeing each spring. It grows in acid soil in woodlands. Soil fungi, *mycorrhizae,* must be present for proper growth.

The flower consists of three petals and three green-brown sepals, but only the large pouch-like petal stands out. The flower stalk is supported at the base by two elliptical leaves with parallel veining. Blossoming time is from April to June.

The genus name, *Cypripedium,* comes from the Greek meaning "Venus slipper," while the species name, *acaule,* means "stemless."

Pink lady's slipper has an unusual adaptation for pollination. The center of the inflated petal has a small slit through which a bee can enter. Bees are attracted to the nectar on the hairs inside. Two holes on the rear wall of the flower serve as exits. Either one will do. As the bee leaves, it rubs against one of the two anthers that are near each exit and carries the pollen to another flower. The stigma along with the anthers are attached to a column that lies between the two exits. An incoming bee will often come in contact with the stigma and deliver the pollen it has brought from another flower.

Fringed Polygala

(Polygala paucifolia)
Milkwort family Size: 3-6″

At first glance, this small delicate flower, sometimes called gay wings, resembles an orchid when spotted in its moist woodland habitat. Closer examination will reveal that this is a member of the milkwort family. The flower, which blossoms from April to July, is an interesting one to examine closely. It has five sepals, two of these are larger, wing-like and colored. The three petals are somewhat united into a tube with the lower petal being boat-like and fringed. This may help to attract insects as well as serving as a landing platform. The color of the flower may vary from pink to magenta. On rare occasions it is white. The upper, ovate leaves are bunched together near the top of the plant. Its lower leaves are few and scale-like.

Polygala is a Greek word meaning "much milk." This is based on an old superstition that cows that eat this plant produced more milk. The species name *"paucifolia"* merely means "few leaved."

Fringed polygala has a second kind of fertile flower that grows on its underground rhizome and never opens. These are cleistogamous flowers. Cleistogamous means "hidden marriage". These flowers are the plant's back-up system in the event that the more colorful flowers do not become pollinated. Most violets and hog peanuts also have this type of flower.

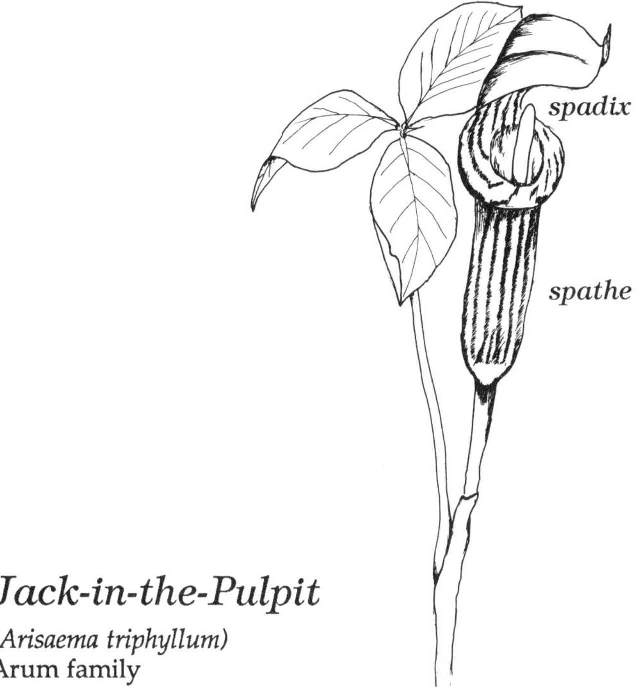

Jack-in-the-Pulpit

(Arisaema triphyllum)
Arum family

Size: 1-2′

This handsome spring flower is easily recognized by its unique appearance. It usually has one or two leaves, each having three leaflets. The unusual looking flower consists of a hooded spathe (pulpit), and a white, finger-like spadix (Jack). Because some individual plants may vary in the color and striping of the spathe, some botanists prefer to list these plants as separate species.

The reproductive parts of Jack-in-the-pulpit are at the base of the spadix. The sexes are usually on separate plants, but can be present in the same plant. Cross-pollination is by fungus gnats and some small beetles.

Jack-in-the-pulpit prefers a moist woodland but will grow on drier sites. It usually blossoms from May to early June.

Each year, depending on growing conditions, this plant must determine if it is to be male or female. A large plant is usually female. This is because it must have sufficient energy to produce fruit, which is a red, berry-like cluster. The same plant, however, may develop into a smaller male plant the following year due to changes in the environment. It may not even reproduce.

The Indians called this plant "Indian turnip" because they ate the turnip-like underground corm after thorough boiling or drying. All parts of the plant are toxic if eaten raw.

Marsh Marigold

(Caltha palustris)
Buttercup family

Size: 8-14″

Marsh marigolds can be found blossoming from April to June in marshes, swamps and along the edge of streams. The sight of clumps of the waxy yellow flowers of marsh marigold in the springtime can be a spectacular sight. However, although they are called marigolds, or sometimes cowslips, they are actually large buttercups.

Marsh marigolds belong to the *Ranunculaceae*, or buttercup family, which includes other spring flowers such as hepatica, wood anemone and early meadow rue. The family name means "little frog," perhaps because close examination of the seeds of this family, reveal that each one looks like a small frog, or merely that many of this family are found in wet environments.

The scientific name *Caltha palustris* means "marsh cup" in Latin. The flowers are cup-shaped with five to nine colored sepals. There are no petals. The stamens and pistils are very numerous, which is a primitive characteristic among flowers. Pollination is carried out mainly by flower flies.

The shiny leaves of marsh marigold are quite large and vary from heart to kidney-shaped. They are attached to stems that are hollow and succulent.

The young leaves of marsh marigold with their stalks removed make an excellent potherb, but they must be cooked thoroughly because all members of the buttercup family can be poisonous if eaten raw.

Chapter 2

Plants In The Spring Woods

Bloodroot - *Sanguinaria canadensis*
Cinnamon Fern - *Osmunda cinnamomea*
Cut-leaved Toothwort - *Dentaria laciniata*
Dutchman's Breeches - *Dicentra cucullaria*
Dwarf Ginseng - *Panax trifolius*
Early Meadow Rue - *Thalictrum dioicum*
Fringed Polygala - *Polygala paucifolia*
Haircap Moss - *Polytrichum commune*
Hepatica - *Hepatica americana*
Highbush Blueberry - *Vaccinium corymbosum*
Jack-in-the-Pulpit - *Arisaema triphyllum*
Marsh Marigold - *Caltha palustris*
May Apple - *Podophyllum peltatum*
Pink Lady's Slipper - *Cypripedium acaule*
Red Trillium - *Trillium erectum*
Rue Anemone - *Anemonella thalictroides*
Sensitive Fern - *Onoclea sensibilis*
Shadbush - *Amelanchier* sp.
Spring Beauty - *Claytonia virginica*
Squirrel Corn - *Dicentra canadensis*
Trout Lily - *Erythronium americanum*
Violets - *Viola* spp.
Wild Geranium - *Geranium maculatum*
Wild Ginger - *Asarum canadense*
Wood Anemone - *Anemone quinquefolia*
Yellow Rocket - *Barbarea vulgaris*

3

Flowers of Street Trees

EACH YEAR WHEN I see a reddish haze in the distance of a nearby swamp, I know that spring is on its way and the red maple is beginning to flower. Shortly afterwards, crimson and yellow splashes of color appear on the maple in my backyard. This early flowering of the red maple seems to be the signal for a parade of intriguing but inconspicuous flowers to burst forth on many of our common street trees. There are a few street trees, such as the silver maple, which blossoms earlier, but their flowers are less colorful and are likely to go unnoticed.

The blooming time of most of our tree flowers is brief, and even some of the small colorful tree flowers may escape our view unless we observe them closely. They are quite unlike some of the more showy or flamboyant flowering trees which adorn our street and lawns.

A commonly held belief is that trees do not produce flowers; however, they do. In fact, some of these flowers are among the most colorful and exotic blossoms found in nature. Many tree flowers can be seen with the naked eye, but others need to be studied with the aid of a hand lens.

One of the reasons we tend to overlook tree flowers is that they are often on branches above eye level, and beyond our reach; however, if we can obtain a sprig of flowers from the tree to examine with a hand lens, it often leaves a lasting impression.

The buds that produce the tree flower do not form in the spring. What we see is the expanding and opening of those buds that were formed long before cold weather set in the previous season. Every year, I look forward to the time when the buds pop and the tiny flowers appear.

Everyone, no matter how limited their knowledge, can experience the joy of witnessing this brief but memorable show. Take a friend for a walk on a regular look around your neighborhood over a period of several weeks. You will discover that each species has its own time schedule for blossoming that may vary somewhat from year to year depending on the weather conditions. Some flowering trees have blossoms which are borne in groups or clusters while others produce single flowers. Each type of tree has its own special appeal.

Trees like the catalpa, the horse chestnut and the basswood are completely without foliage when trees such as the red maple are sending forth new flowers and leaves. Catalpa and horse chestnut do not flower until late spring and the leaves appear before the blossom. Leaves are obstacles to wind-pollinated trees like the maples, birches and oaks. This accounts for the flowers appearing at or before the time the leaves begin their rapid growth. Catalpa and horse chestnut have showy flowers which are insect-pollinated and leaves do not interfere with this process.

Basswood, with its sweet smelling blossoms, doesn't usually flower until sometime in July. Because of the odor from the blossom, which attracts honeybees, it is often called "the bee tree."

I usually go for daily walks in my neighborhood. After the red maple has blossomed I watch for the dangling bell-like flower of the sugar maple and the handsome bloom of the Norway maple which seems to explode from its swollen buds into a yellow-green spray. The flowers of the white birch, in contrast, are slender, tassel-like catkins that appear about the same time as the sugar and Norway maple.

The cottonwood and quaking aspen also bear catkins along with the oaks and hickories. With the exception of the red maple, all these as previously mentioned are wind-pollinated. The red maple with its colorful flower is pollinated by both wind and insects. The sassafras may vary considerably in its blossoming time. This may be as early as April or as late as June. The leaves and flowers emerge at approximately the same time.

The white pine produces what some people like to call "sulfur showers." This is because of the huge amount of pollen that is produced by this wind-pollinated tree.

There may be some question as to whether pines have true flowers because their seeds are protected only by the scales of a cone and they become exposed when the cone opens. Flowering trees with true flowers have their seeds enclosed inside an ovary. In spite of these technical differences, pines will be discussed here with the flowering trees.

Although many people are familiar with the flowering dogwood, few are aware of the structure of the flowers. The true flower is actually a cluster of tiny complete flowers surrounded by the showy pink or sometimes white bracts that are often mistaken for petals.

Flowering dogwoods have buds that are of two kinds, mosque-like flower buds, and small, more pointed leaf buds. The flower buds form on the tip of the branches and open first. The leaf buds are located right below them and open later. Many species of trees have mixed buds. These buds produce the flower and leaves from the same bud.

Several of the trees that I have mentioned belong to a plant family which consists solely of trees, such as the different kinds of maple; while others belong to families which include shrubs, vines and other small flowering plants.

The black cherry and shadbush, are two species of trees that belong to the rose family, but so do flowers like the wild strawberry and the common cinquefoil that you may find growing on your lawn.

From early times, man has depended on trees for fuel, shelter, medicine and for helping with some of his spiritual needs. Trees are among the oldest living things on earth. Some of the bristle cone pines of California have been found to be over 4000 years old. Only the creosote bush of the American desert is older. Some of these plants have been growing and expanding at one site for nearly 12,000 years.

Trees are expressions of the splendor and mystery of life. Over the years, man has drawn a great deal of inspiration from trees, but there is a danger that he may not be able to do this so much in the future. At present, we tend to think of trees largely because of their practical usefulness. There is concern about our rapid use of forests and how this will affect us in years to come.

If we can become more familiar with trees in our immediate surroundings and appreciate them in an aesthetic way, we may continue to draw satisfaction and inspiration from them. Observing tree flowers and learning the value of neighborhood trees in attracting wildlife, particularly birds, is one way to maintain emotional ties with our environment which we are in danger of losing.

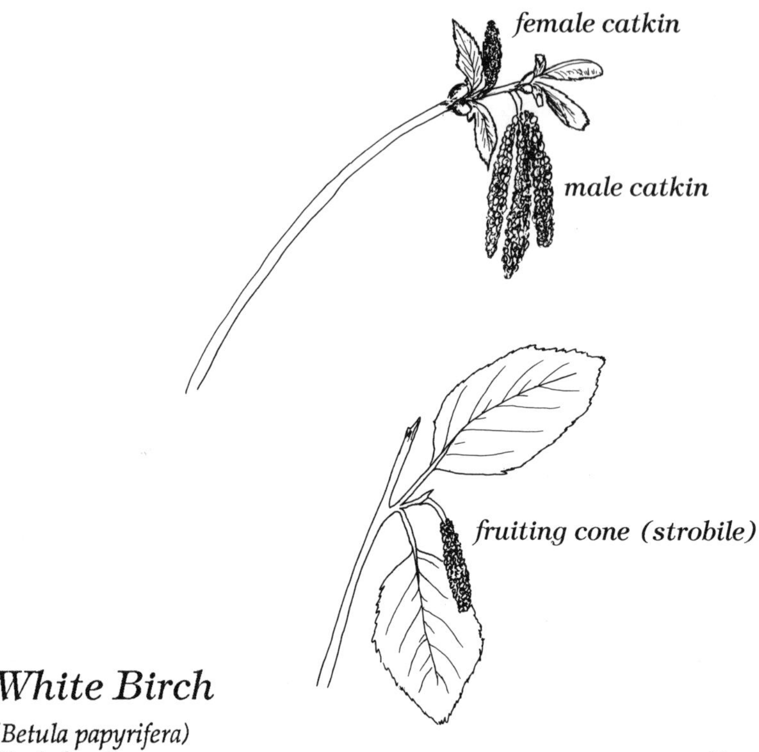

White Birch

(Betula papyrifera)
Birch family

Size: 70-80′

White birch is one of our most popular native ornamental trees because of its beautiful bark and graceful form.

It has pendant tassels which are called catkins, apparently because of their imagined resemblance to a cat's tail. They are actually flower clusters as opposed to single flowers and they are wind-pollinated.

The beauty of the white birch with its striking predominately white bark, lies both in the total impression of the tree, that may have hundreds of male catkins dangling downward, and in the male and female flower clusters themselves. The drooping male catkins are three to four inches long, slender with reddish brown scales. There is a single female catkin that is short, slender and up to 1$^1/_2$ inches in length and curves upward above the male catkins. The flowers usually appear in May just before the leaves. Staminate catkins are in groups of two or three and start growing in the spring when they are about one inch in length. They wintered through in this form.

The cone-like fruit is about 1$^1/_2$ inches long and hang downward, producing small winged seeds. They are food for birds such as the American goldfinch, blue jay, black-capped chickadee and tufted titmouse. These birds also use white birch for cover. This is the traditional tree from which birch bark canoes are made.

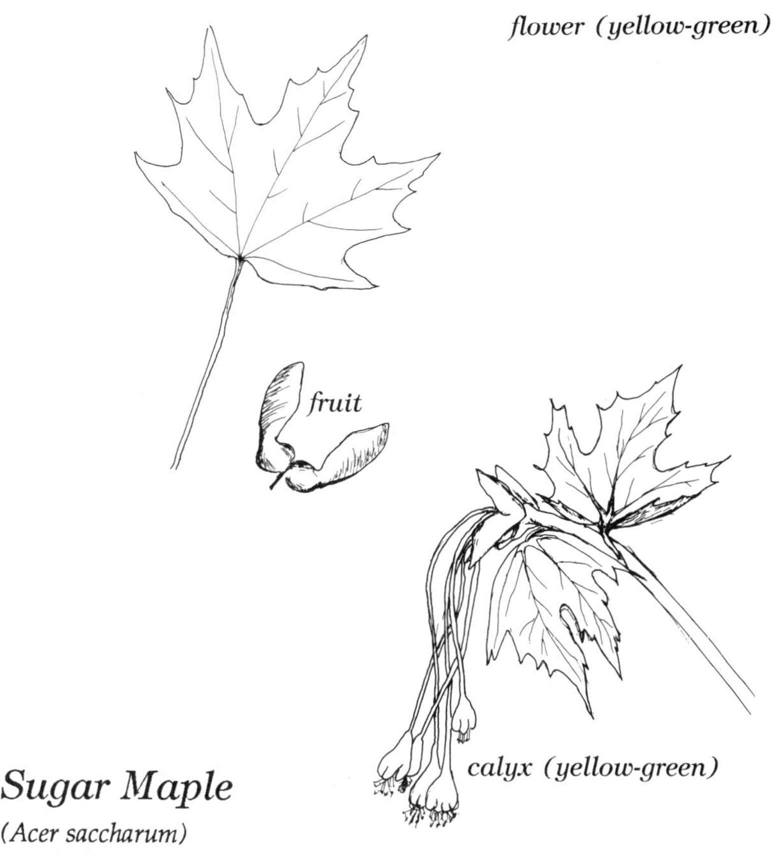

Sugar Maple

(Acer saccharum)
Maple family

Size: 40-60′

Sugar maple is an ornamental tree because of its attractive round-topped crown, shade-value and colorful autumn leaves. The sap of this tree is the source of the familiar maple syrup and sugar.

Sugar maple may be easily recognized by its pendant tassel-like flower clusters. The leaves and blossoms appear together in late April or May. There is a five-lobed calyx, but no petals. The two types of flowers, pistillate and staminate, may be borne in the same cluster or in separate clusters, or sometimes the two kinds are on different trees. If on the same tree, the seed bearing tassels are at the tips of the twigs and those bearing pollen are along the side.

While the staminate flower shows little or no trace of a pistil, the pistillate flower possesses its complement of stamens, usually about eight, but these do not produce viable pollen. The mature seeds are not completely formed until autumn and resemble rabbit ears, the wings being quite close to each other.

The rose-breasted grosbeak feeds on the seeds and uses the tree for cover and nesting, as do the red-eyed vireo and American goldfinch.

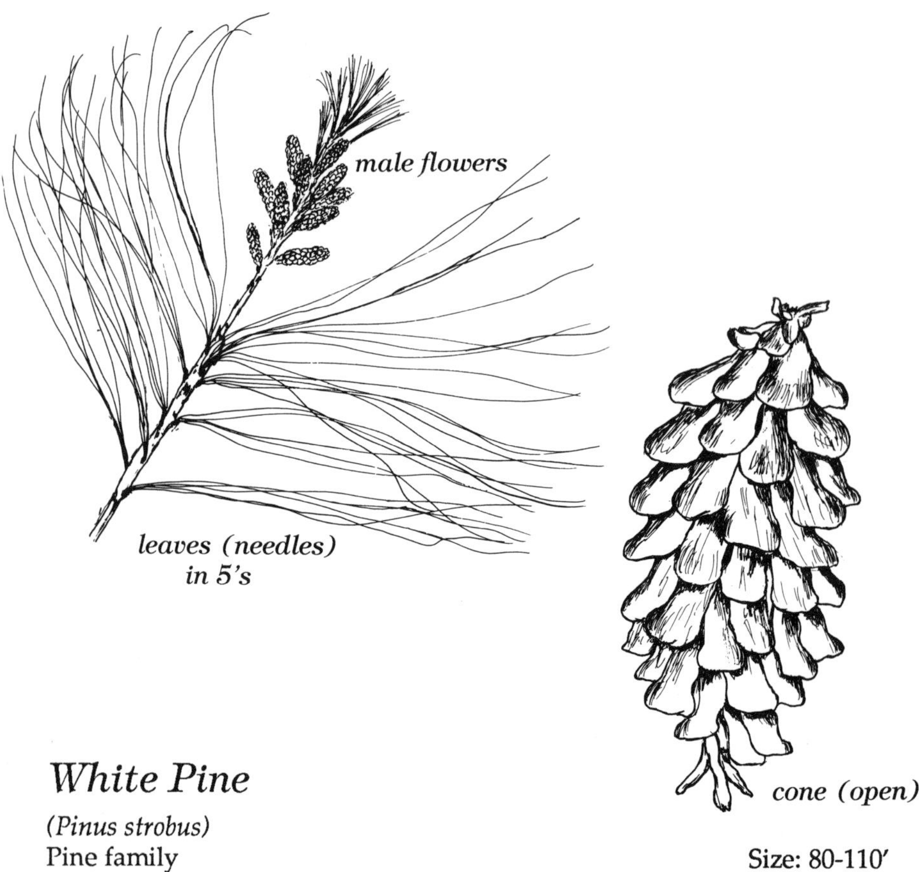

White Pine

(Pinus strobus)
Pine family

Size: 80-110′

The stately and graceful eastern white pine is a popular ornamental at all stages of its growth. Historically, it played an important role in providing the early settlers with ship masts and wood for their homes.

White pine flowers in June and two years are required for the development of the scaled cone. The male or staminate flowers are widely distributed on the tree in clusters. Each flower in the cluster contains many plump stamens arranged in a circle around a central axis. Each stamen has two pollen sacs which, at the right time, will split and a virtual shower of pollen will be distributed by the wind.

The female or pistillate flowers are formed near the tip of the new twigs, usually on the higher branches. There are generally only one or two forming on each twig. They are green at first and become reddish when they are pollinated. Following this, a cone begins to develop. At the end of the first year, it has become nearly an inch long. The second year, the cones expand to a length of five or six inches, and the scales become dry and split open. The winged seeds drop out and are dispersed by the wind. The seeds are the food for birds such as the black-capped chickadee, tufted titmouse and cardinal.

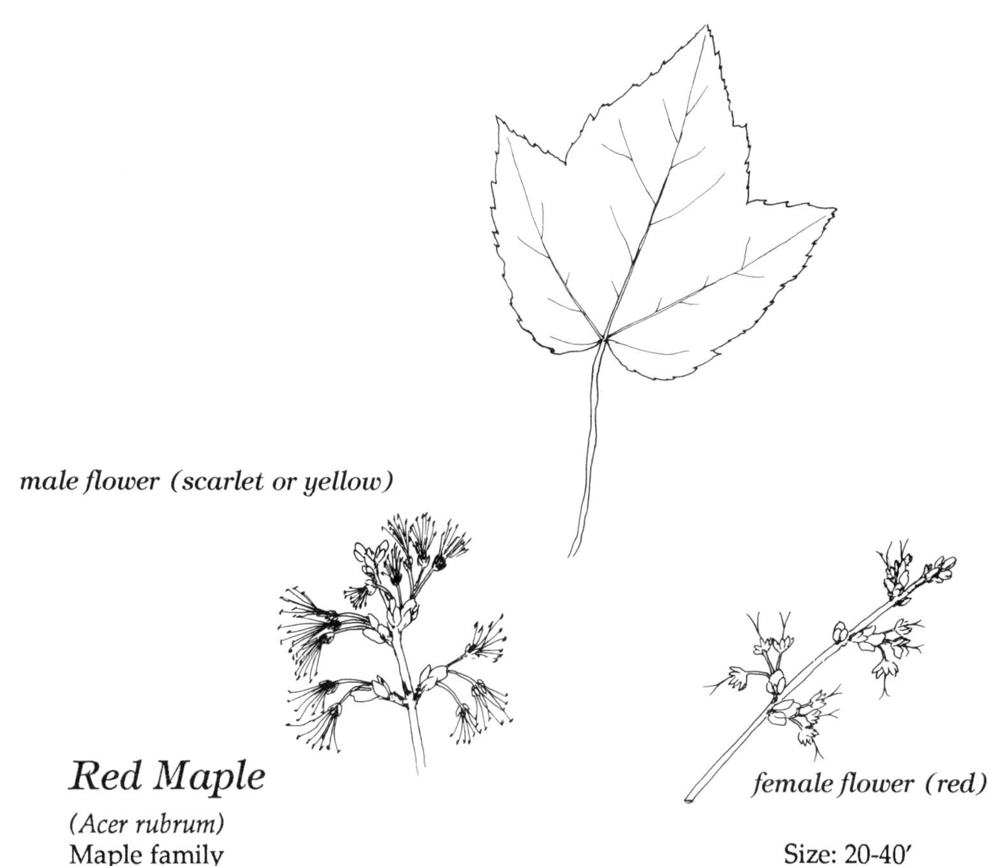

male flower (scarlet or yellow)

female flower (red)

Red Maple

(Acer rubrum)
Maple family

Size: 20-40′

Red maple is a native tree, often found in swamps and low ground as one of the dominant trees. Along streets and in yards it adds a touch of color when its flowers first blossom in the spring. It is well named because there are many things about it that are red. The flowers are red, its leaf stalks are often red, the winged seeds are red and as the leaves turn in September, they become red. This gives a swamp a rosy appearance from a distance.

The buds of red maple burst open in February and the blossoming period continues to May. The flowers may be scarlet or yellow and they appear before the leaves. they are arranged in pairs of opposite clusters, each on a long thin stalk. The stamens may be either scarlet or yellow and usually vary from five to eight in number and are quite long. The female flowers have stamens that are either aborted or are rather short.

Fertilization occurs rapidly and small red, wing-like seeds, characteristic of maples, soon appear. This fruiting period is from March to July.

Cardinals eat the seeds, yellow bellied sapsuckers get sap from the tree and the goldfinch and robin use the tree for cover and nesting. The early settlers made ink and brown and black dyes from extract of the bark.

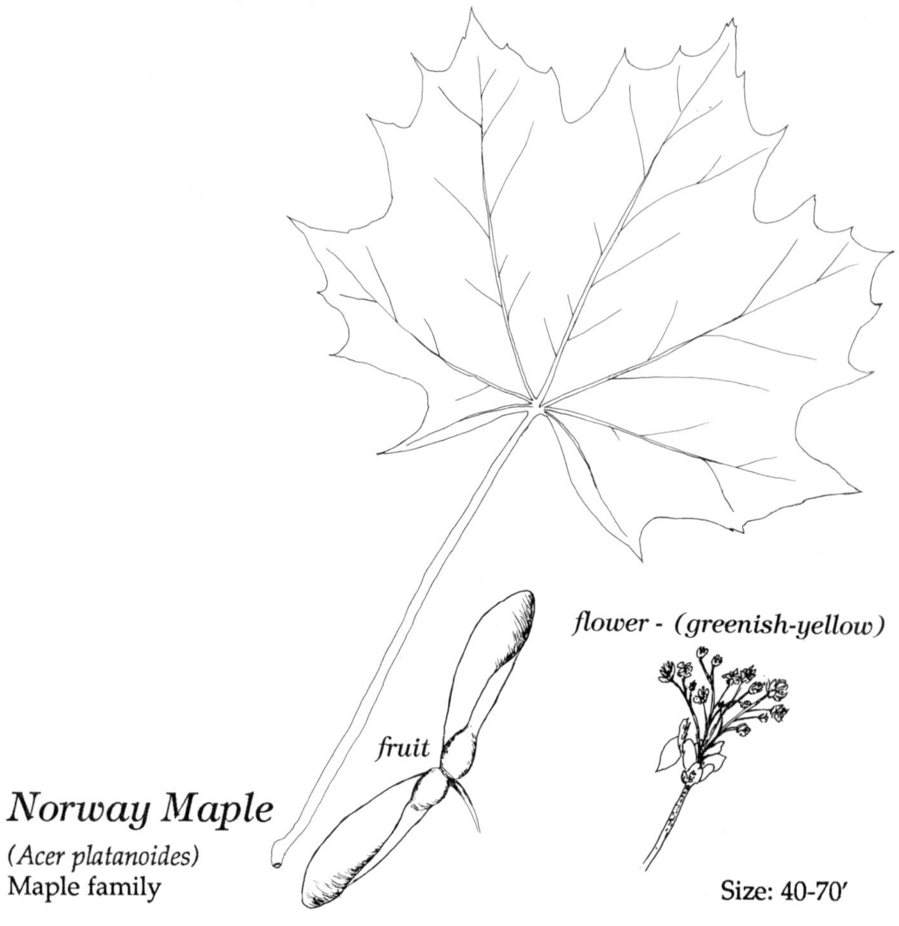

Norway Maple

(Acer platanoides)
Maple family

Size: 40-70′

One of the most common street trees is the Norway maple, a tree imported originally from Europe. It blossoms from April to June. Norway maple can grow in nutrient-poor soil and is more tolerant of air pollutants than native species. It has dense foliage that persists in the fall and milky sap in its leaf stalks. Horticulturists have developed many interesting forms of this hardy tree.

The flowers of the Norway maple are yellowish-green and they appear in erect, smooth spherical clusters, tight at first, and then almost seem to explode into more open, spray-like clusters, like a "Roman candle."

The male and female flower clusters are very similar, but in the male the pistil may be present, although it is only a rudiment. In the female flower cluster, the stamens may occur in the characteristic number, but they never reach full length, and pollen is not discharged because the anthers do not open. The pistil is the distinguishing feature of the flower, appearing in the central position. The leaves develop at about the same time as the flower clusters.

The seeds that form are paired. They mature in September with wings widely spread. Robins and goldfinches feed on the seeds of Norway maple and use it also for cover and nesting, as does the northern oriole.

Chapter 3

Flowers of Street Trees

American Basswood - *Tilia americana*
Black Cherry - *Prunus serotina*
Catalpa - *Catalpa bignonioides*
Eastern Cottonwood - *Populus deltoides*
Eastern White Pine - *Pinus strobus*
Flowering Dogwood - *Cornus florida*
Hickories - *Carya* spp.
Horse Chestnut - *Aesculus hippocastanum*
Norway Maple - *Acer platanoides*
Oaks - *Quercus* spp.
Quaking Aspen - *Populus tremuloides*
Red Maple - *Acer rubrum*
Sassafras - *Sassafras albidum*
Shadbush - *Amelanchier sp.*
Sugar Maple - *Acer saccharum*
White Birch - *Betula papyrifera*

4

The Lowly Violet

MOST PEOPLE ARE able to recognize the modest violet, often part of the array of early spring flowers. Perhaps less notable than their familiarity as a flower is their tremendous diversity in color and parts. Many of these differences constitute the appeal of this group of flowers.

Violets, over time, have become a word for a color. This is something it has in common with the rose. However, not all violets are violet in color. They may be white, cream, yellow, purple, many shades of blue and even green-brown. Some violets are fragrant; there are others that are not.

Violets look like small pansies. There are two types based on the absence or presence of a stem. Stemless violets have leaves and flowers arising on separate stalks, while the other type, the "stemmed" violets, have an erect stem, bearing both leaves and flowers.

My enthusiasm for these diminutive charmers become enhanced with my first sight of the beautiful birdfoot violet, I found these growing in patches in a sandy area. Discovery of the unusual green violet along a stream in a moist woods and a white, yellow-throated Canada violet in August along a roadside in Vermont, added fuel to the fire. I began to ask myself questions.

Telling the difference between one species of violet and another may seem baffling at first, particularly concerning the stemless blue violets. This is where the use of a hand lens is very helpful. It is a valuable aid in appreciating the beauty of small details and in thus establishing the identity of a particular species. There is the beauty when we see the whole violet plant with its leaves, stem and flower, but there is also the beauty of the fine details.

Violets occur in a variety of habitats so it is necessary to keep this in mind when taking a walk to find them. However, the study of violets can start in your own backyard, by considering the garden pansy, which is a violet *(Viola tricolor,* var. *hortensis).* It is an offshoot of the irrepressible Johnny-jump-up (*Viola tricolor*). These pop up in lawns and grow along roadsides and fields. Years of experimentation with hybridization by horticulturists have produced the present day pansy with its many strains.

The name of our garden pansy comes from the word "pensee" meaning thought. The French originally gave this name to the Johnny-jump-up, believed to act as a love potion. This was undoubtedly why another of the common names of the Johnny-jump-up was heartsease.

Also growing in your yard you may find the common blue violet. Other favorite habitats for this flower are roadsides and meadows.

Going farther afield, we can look for some other wild violets. Late April, or early May, is an ideal time for our search. As with the other spring flowers, generally the larger the area we explore, the greater the chance of finding varying habitats and a variety of species.

I have a favorite place where I like to look for violets. It is comprised of several woodland trails including a path around a lake. Here, in the rich moist woods we can look for several different species. Some of the common violets that grow here are the round-leaved violet, an early bloomer, the dog violet, usually growing in grassy areas along the trail, the sweet white violet and the smooth yellow violet. The downy yellow violet may also be here, as well as on drier sites.

Nearer the lake, the marsh blue violet and northern white violet grow in wet mucky soil. The northern white violet is sometimes even found growing in shallow water.

Most of our common violets will be found growing in rather shady conditions. Many of them grow where it is either moist or wet. A few species of violets, such as the birdfoot violet and the ovate-leaved violet, prefer dry sunny areas where the soil is poor.

Once we have discovered a number of species, it is then possible to see many of the ways one violet differs from another. It is helpful to know something more about the general characteristics of the violet family and their uses before we go much further with identification.

This family, for the most part, consists of low-growing plants. Incidentally, there are some tropical species of violets that reach tree size. Violet flowers have five petals. The lowest petal is often wider and heavily veined. It serves as a landing platform for insects and extends backward into a spur, which contains a large horn-shaped nectar gland. The lateral petals are often hairy or bearded. The variations in these petals are often the key to telling one species of violet from another.

Each flower contains five stamens which closely surround a single pistil. The two lower stamens extend into the spur and contain nectaries. These are bodies that secrete nectar. The solitary pistil is distinctive, usually having a club-shaped style with a head. Its appearance varies somewhat depending on the species. In some violets, the head of the style bears a beak. The green violet (*Hybanthus concolor*) is a plant that would never be suspected of being a violet except for the club-like pistil. These flowers are small, green and they droop singly from the axils of the leaf.

The fruit of violets is in the form of a capsule that explodes to disperse its seeds. There is considerable variation in species of violets as to the color of their fruit and seeds.

Reproduction in violets is performed in two ways by flowers that are quite different in appearance. The showy spring flowers are quite obvious and are cross-pollinated by insects in search of nectar. During the summer months, petal-less, closed flowers are present on most species. These are often borne on short stalks near the ground. They are called cleistogamous flowers, meaning "hidden marriage." These flowers are equipped for self-pollination. This is a remarkable back-up system to ensure survival in the event that such things as rain, cold weather or too much shade interfere with cross-pollination in the petaled flowers. Fringed polygala and hog-peanut have similar flowers. Many violets are able to produce showy flowers again in the fall. One of the requirements is that there must be the same number of hours of darkness as was present in the spring.

Violets, with the aid of insects, are constantly cross-breeding in the wild. The violet may be a hybrid of a hybrid. This means that the traits of four species of violets could be involved. There are about 300 species of violets worldwide. Approximately 80 species grow in the United States.

Down through the ages, violets have had a number of significant uses, culinary and medicinal. During the days of Ancient Greece, a violet concoction was a method of producing sleep for insomniacs and to give strength to those who had heart trouble. Women in pre-Roman Britain used a medicine made from violets and goat's milk to help their complexions.

Violet leaves contain high amounts of vitamin A and C; more than that present in oranges and spinach. This has made them a favorite for a potherb. They may be cut up and used in a salad. The leaves may also be used to thicken soup. Flowers of violets can be candied or they may be used to prepare jams, jellies or syrups.

Coming upon a clump of violets as a harbinger of spring has its own special appeal. However, discovering the intricacies of this lovely flower on closer inspection and learning of its many uses may open a whole new world to you.

Birdfoot Violet

(Viola pedata)
Violet family

Size: 4-10″

Birdfoot violet has been called the loveliest of the stemless wild violets. The flower color in this species varies. Two common varieties include one with uniformly colored petals and another where the upper petals may be deep purple and the three flower petals are paler.

The five petals of the lower of this species are all beardless. The blossoms are larger than most violets, sometimes being over an inch wide. It has five orange stamens that stand out. The upper two petals are reflexed backwards.

The birdfoot violet lacks self-fertilizing (cleistogamous) flowers and must depend on cross-pollination for the production of seeds. These must all come from petaled flowers early in the season. The fruit capsule is green.

This species is one of the few violets that thrive in open dry sites, such as sandy fields, wood openings and rocky banks. Birdfoot violet blooms from March to June and may flower again in the fall if conditions are right.

The leaves of this violet are quite unlike those of other violets. The species name *pedata,* means foot-like, referring to the manner in which the deeply cut leaves fan out to resemble a bird's foot. Birdfoot violet was used by the Indians to make a medicine thought to be helpful in dealing with lung and urinary problems.

Common Blue Violet

(Viola papilionacea)
Violet family — Size: 3-8″

This is probably the wild violet that we see most often. Its habitats include borders of woodland, meadows, roadsides and dooryards. This species is one of the many blue violets with flowers and leaves on separate stalks, both originating directly from a thick and fleshy rootstalk.

The species name, *papilionacea,* meaning "butterfly-like," refers to the appearance of the corolla. Its colorful flower varies from a bluish-purple to light blue, depending on soil conditions and the amount of light. The center is white. This species also has a white form, called the "Confederate violet." The flower is whitish or pale gray with blue veins and is often used as a cultivated flower.

The flowers of the common blue violet extend only slightly above the heart-shaped leaves. In the similar common marsh blue violet, the flowers rise well above the leaves. There are five petals to the blossom with only the two side petals being bearded. The beard, or hairs, are mostly with slender tips rather than with tips that are knobbed. Its three lower petals are all deeply veined.

The self-fertilizing (cleistogamous) flowers are on stalks that are low to the ground. Its fruiting capsule is usually mottled with purple, while the seeds generally have purple dots.

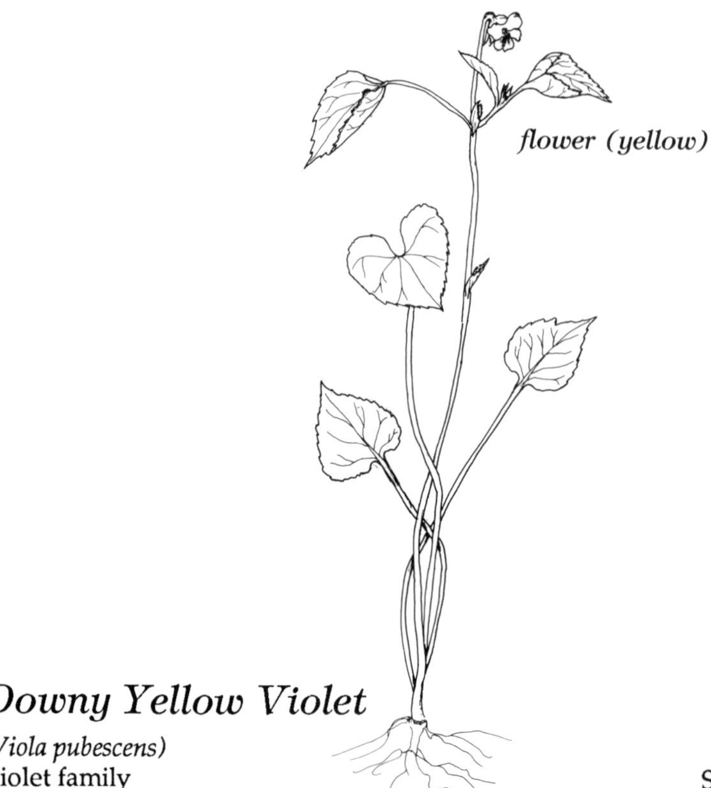

Downy Yellow Violet

(Viola pubescens)
Violet family

Size: 6-16″

The distinctive characteristics of this stemmed yellow violet is the presence of soft downy hairs on its leaves and stem.

This is rather a tall violet varying in height from 6 to 12 inches. The leaves are characteristically pear-shaped, pale green and velvety. In this species, there is usually an absence of leaves at the base of the plant, but occasionally one leaf is present. Below the upper leaves, stipules are present. These are leaf-like and slightly toothed.

The downy yellow violet blossoms from April to May. It is often found growing in dry, rich deciduous woods, but may grow on moist sites. Only a few lemon-yellow flowers develop in the axil of the leaves. The two lateral petals are bearded and the three lower ones are veined with purple. The lowest petal is spurred and shorter than the others. Its fruit capsule may vary from wooly to smooth, the seeds being pale brown.

The smooth yellow violet (*Viola pensylvanica*) is similar to, and sometimes hybridizes with the downy yellow violet. Typically, if it is not a hybrid, its leaves and stems will be hairless or nearly so. There will be one to five basal leaves and the stipules will lack teeth.

It is said that the American Indian applied bruised leaves of yellow violets for treatment of boils and painful swellings.

Marsh Blue Violet

(Viola cucullata)
Violet family

Size: 5-10″

This stemless blue violet has flowers on long stalks that extend much beyond the leaves. It is sometimes called the long-stemmed marsh violet. It is commonly found in wet or moist areas such as went meadows, swamps and bogs, in sun and in shade. Sometimes it spreads to drier ground and may appear faded in color.

The flower and leaves arise from a thick and sometimes short rootstalk. This branches to form colonies of this species. its flowers blossom from April to June and are usually violet blue. They are often darker blue at the throat, but may be white with darker veins.

The side petals are bearded, with the ends of the hairs being knobbed or club-shaped. A hand lens is needed to see this characteristic. The common blue violet (viola papilionacae), also has side petals that are bearded, but the hairs are not knobbed.

The lowest petal of the marsh violet is smooth, spurred and veined. This petal is somewhat shorter than those on the side. The leaves are variable, but in general are heart-shaped with either blunt or pointed teeth. Cross pollination, if it occurs, is carried out mostly by bumblebees. Its self-fertilizing (cleistogamoous) flowers are on long erect stalks. The fruiting capsule is green and the seeds are nearly black.

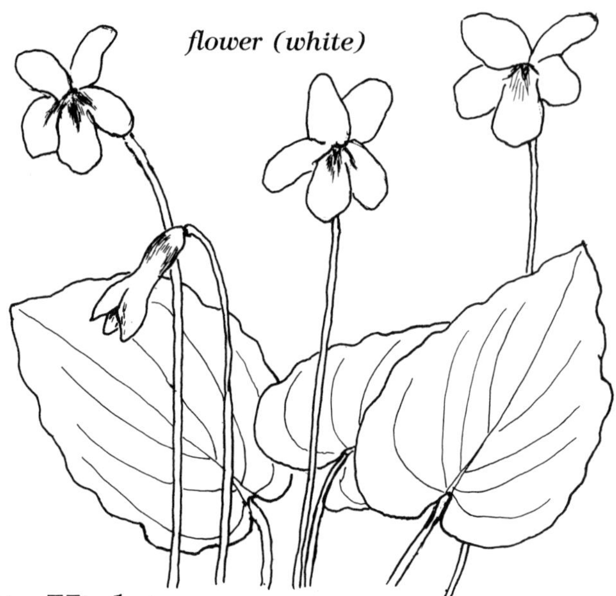

Sweet White Violet

(Viola blanda)
Violet family

Size: 3-5″

This fragrant flower is one of our smallest violets with flowers only about a half inch wide. It likes cool, shady, moist places such as ravines and rich woods.

The blossom is five-petaled, beardless, with the two upper petals being longer and narrower than most violets. They are also bent backward or twisted. The lower petals are purple-veined. Its fruit is a purple capsule producing brown seeds.

This stemless violet develops separate leaves and flowers from a slender underground stem. Later in the season, the plant produces runners, or stolons, from which new plants develop. These runners are branched, making it possible for the plant to spread out and claim a larger territory.

The flowering time for the sweet white violet is from April to May. This is later than some of the earliest violets.

Cross pollination, if it occurs, is carried out by honey bees and certain other small bees. They are probably drawn to the plant because of its fragrance. Its self-fertilizing (cleistogamous) flowers develop on stalks that lie near or on the ground.

The leaves of this species are dark green, heart-shaped and sharp pointed at the top. The base of the leaf is rather deeply cut. The northern white violet (*Viola pallens*) is a similar species, but it has a red stem and its seed capsule is green.

Chapter 4

Violets

Birdfoot Violet - *Viola pedata*
Canada Violet - *Viola canadensis*
Common Blue Violet - *Viola papilionacea*
Dog Violet - *Viola conspersa*
Downy Yellow Violet - *Viola pubescens*
Garden Pansy - *Viola tricolor,* var. *hortensis*
Green Violet - *Hybanthus concolor*
Johnny-jump-up - *Viola tricolor*
Marsh Blue Violet - *Viola cucullata*
Northern White Violet - *Viola pallens*
Ovate-leaved Violet - *Viola fimbriatula*
Round-leaved Violet - *Viola rotundifolia*
Smooth Yellow Violet - *Viola pensylvanica*
Sweet White Violet - *Viola blanda*

5

Lawn Plants

THE FAMOUS NATURALIST Louis Agassiz once said, "I traveled all summer, and I only got half way across my backyard." We need not travel great distances to discover the beauty and excitement of the natural world . . . it lies close at hand - right outside your window in fact.

History reveals that many plants we do not consider wild were first brought to this country by the early settlers to plant in their gardens. They were chosen because of their beauty and usefulness. With the passage of time, many of these plants escaped from the formal garden and became established on lawns and along roadsides. Many weeds have interesting stories to tell as we will see when later on we take a closer look at five of those you might find on your lawn.

No two lawns will be exactly alike although they will be similar. Each lawn is a special environment (or ecosystem), which means that the physical conditions present are always at least slightly different.

Some lawns will have different soil conditions from others. There will be varying amounts of moisture and shade. Lawns with trees will attract different animals than those without trees, and the population of birds squirrels and various kinds of insects will depend on the types of trees that are present. In addition, the bases of some of these trees will have special micro-climates that will be particularly favorable for certain types of plants. All these factors give rise to different inter-relationships.

Examine closely the physical and biological features present along the border of your lawn. These could be shrubs, trees, a sidewalk, a gutter, or a vacant lot containing grass that is not mowed. All these factors influence the amount of sun and the temperature as well as the plant and animal life.

Your lawn may help you to understand other parts of the world, some of which you may never visit, because in many ways it is similar to deserts, grasslands and forests that you may have read about.

When I get restless for more contact with the natural world, I often take a walk around my yard. I am always amazed at the diversity of plants that are present. I am invariably intrigued by the tiny speedwells. I find two different species growing on my lawn . . . the common speedwell and the thyme-leaved speedwell. I have found corn speedwell and slender speedwell on other lawns. Some of these form solid mats in certain places. These are all small plants, but their beauty is not obvious until observed through a hand lens. The chickweeds are similar with their deeply cut petals, and may be found along with bluets, gill-over-the-ground and moneywort in somewhat shaded spots.

A couple of other favorites are the persistent dandelion with its many tiny ray flowers containing both male and female parts and common plantain, a good example of a plant that we never associate with beauty, but the hand lens tells us otherwise. When its spike-like flower is in blossom, it is a thing of exquisite beauty.

Some of the larger plants that appear on my lawn, especially along the edges, are white campion, curled dock, silvery cinquefoil, kidney-leaved buttercup and Indian tobacco. All of these are broadleaf plants, often found growing with lawn grasses. One Connecticut botanist recorded some 50 species in a typical lawn that had not been treated with herbicides. Many of these species, if not all, are commonly called weeds. However, much depends on your point of view. Ralph Waldo Emerson illustrated this point quite well in 1878, with this quotation appearing in the magazine called Fortune of the Republic: "What is a weed? A plant whose virtues have not yet been discovered."

In order to become familiar with some of these common plants, a good wildflower guide such as Peterson's Wildflower Guide or Newcomb's Wildflower Guide are a valuable aid. Those who like to use keys may prefer Newcomb's, because this guide has a simple key that requires no past botanical knowledge.

Once the plant's identity is known, you may want to expand your knowledge and understanding by asking yourself such questions as: Is it edible? Does it presently have, or has it ever had, any medicinal use? How did it get its name? Does it have any other uses? What are some of its adaptations? Is it poisonous? Enthusiasm is contagious. Your eagerness to know more about these plants may well spark an interest in others.

Lawns evolved long before the days of professional gardeners and landscape architects. They were used to keep domesticated animals tethered close to home. The excretions of the animals fertilized the soil and their stamping hooves allowed only grasses to survive. They kept the grasses short by their constant grazing. All that has passed and the family cow has been replaced by herbicides, chemical fertilizers and the power mower. Fill has replaced the original top soil. Lawns now serve a different purpose.

Lawns can be as dull or as interesting as you want to make them. They can be just grass and trees or a natural setting that attracts birds, insects and other animals and adds to your understanding of the environment as well as your pleasure. Have you ever thought of managing your lawn to attract birds and other wildlife? Ecologist Frank Egler suggests that much of the lawn of the average American home can profitably be converted to various plant communities. In this way the actual lawn size is decreased, which is ecologically sound and useful, particularly in suburban, and semi-rural areas.

Bluets

Hedyotis (=Houstonia) caerulea
Madder family

Size: 3-6″

This endearing and delicate little plant was named after William Houston, a Scottish botanist. The species name *caerulea* is Latin for "sky blue." The word bluet is a French word relating to its color. Its other common names include Quaker-ladies and innocence.

Large colonies of this plant are often found beneath mature trees on lawns where they have little competition. They may form small clusters resembling patches of snow. The tiny four-petaled sky-blue or white flowers have a bright yellow center and blossom from April to June.

The flowers are dimorphic, meaning they occur in two forms. The anthers and stigmas are placed in different positions in the two types of flowers. Type A has flowers with a tall stigma protruding from the throat of the flower tube and short anthers placed below. Type B has flowers with a short stigma with long anthers above.

Aided by bees, the tall stigma from type A flower is pollinated from the tall anthers of type B flower, while the short stigma of type B is pollinated by the short anthers of Type A. This adaptation of the flower structure is an ingenious method of insuring cross-pollination.

Common Speedwell

(Veronica officinalis)
Vervain family — Size: 3-10″

The common name of this lovely little flower probably evolved from the charming tradition of handing fresh blue flowers to travelers with the wish "speed well." The species name means "of the shops" referring to apothecary shops and this plant's medicinal uses. It was reported to possess diuretic and astringent properties.

The genus name is believed to come from the Greek words vera "true" and eicon "image." This is connected to a legend dealing with the true image of Christ, left on a towel when a woman wiped the sweat from Christ's face as he was on his way to Calvary. The woman later became Saint Veronica, for whom the flower received its name.

There are about 20 species of speedwell in the East. They are all plants that do well in full sun, but they can grow in spots that are somewhat shady. They are all low, creeping plants. Their flowers are usually blue with generally, a four-parted corolla that is slightly irregular. There are two stamens and one pistil. A number of species came to our shores as stowaways in grain seed and ballast. The lovely lilac-blue or blue-striped white flowers of the common speedwell often go unnoticed. You can look for the blossoms between June and August.

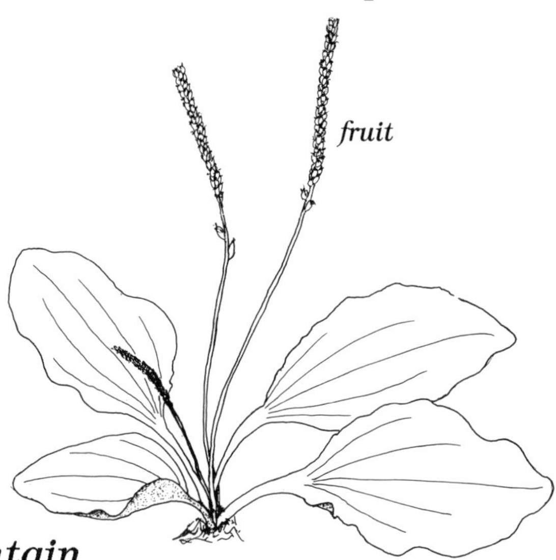

Common Plantain

(Plantago major)
Plantain family

Size: 6-18″

This persistent and widespread plant has a fascinating history. It was considered at one time as a valuable herb and an edible green, as well as being useful as bird seed. It was grown in every monastery garden and also cultivated in botanical gardens. People today appear to be rediscovering some of these uses, which never completely disappeared over the years.

This Eurasian plant was first brought here by the Puritans of New England to be used to help them survive. The seeds were quickly dispersed by birds and the boots of settlers. It was called "white man's foot" by the Indians because it appeared to spring up wherever the settlers moved. This is an odd coincidence because both the generic and common name of this plant come from the Latin word *plantago* meaning "sole of the foot," this name referring to the shape of the leaf. *Major* means "large," which describes its leaves.

Common plantain has tiny flowers massed together in a narrow greenish spike. This rises from a set of broad basal leaves. The pistils mature before the stamen so flowers of one part of the spike may pollinate another part of the same spike with the aid of wind, gravity and insects. The fruit is a capsule that splits at maturity in a circular manner, near the top. The upper part of the capsule, a lid, drops off and seeds are dispersed.

Two other common species of plantain found on lawns are the pale plantain and English plantain. Pale plantain is very similar to common plantain but its leaf stalks have reddish bases. English plantain has narrow leaves that are heavily ribbed.

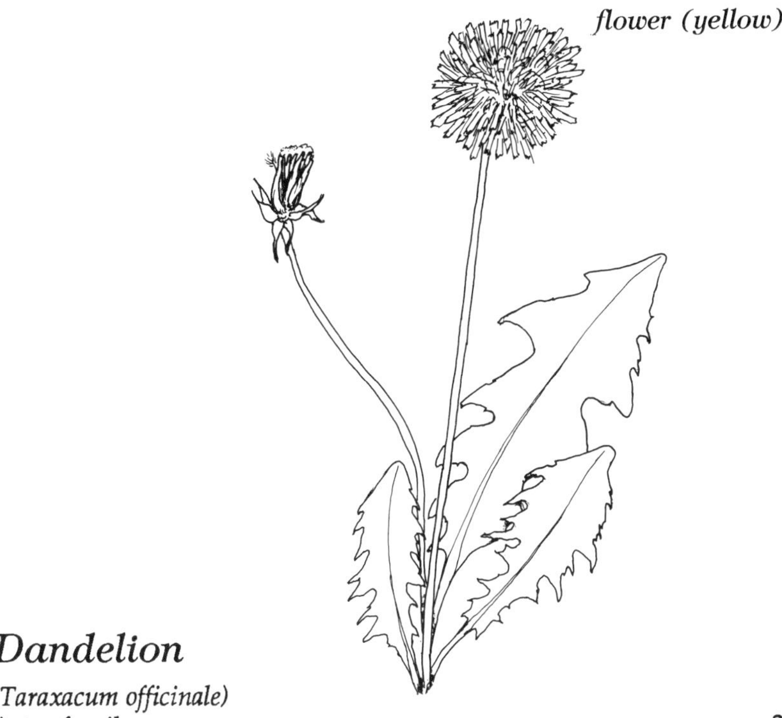

Dandelion

(Taraxacum officinale)
Aster family 2-18″

The common name of this golden flower is a corruption of the French "dents de lion," teeth of a lion, referring to the sharp projections on the margins of the leaves. Each petal comprises one entire flower, a ray flower that has both stamen and pistil. Each strap or ray flower has five teeth at the end indicating a five-parted corolla.

The dandelion produces a great deal of nectar and pollen. This is a rich source for young bees when they are developing in the hive. They require a constant supply of bee bread, which is a mixture of pollen and honey, in order to mature. Bees are then able to pollinate many other flowers. In spite of the fact that dandelions produce pollen, they are sterile. There is no true pollination or fertilization, but the female parts can develop a fruit, and disperse with the wind. You may wish to use a hand lens to see the floral parts.

The early colonists considered the dandelion a common potherb, and the young leaves were used in salads. Its seeds were among those brought to this country for the family garden. The roots have been used as a substitute for coffee, while the flowers can be steeped in water to make a delightful wine.

At present, dandelion roots are used by pharmacists in tonics and liver medicines.

Gill-Over-The-Ground or Ground Ivy

(Glechoma hederacea)
Mint family

Size: to 8″

The leaf color and trailing habit of this little mint are what give the plant its scientific name. The generic name *Glechoma* refers to the gray-green color of the leaves, and the species name *hederacea* means "resembling ivy." The name gill probably comes from the French "guiller" because the leaves were once valued for helping to flavor or ferment beer.

Its square stem, opposite leaves and irregular flowers are all good mint-like characteristics. It has a very attractive, tiny, hooded blue flower that blossoms from March to June, but unlike many mints, this plant is not aromatic.

Gill-over-the-ground is an alien plant that has become naturalized, having been introduced into this country from Europe. It is now found growing in moist, shaded or sunny spots on lawns and along roadsides.

This plant has a history of being associated with witchcraft and magic. For many years, it was believed that any plant growing near it would die.

As far back as the Ancient Greeks, gill-over-the-ground was well known as a medicinal herb. Today, its main claim to medicinal fame is the high vitamin C content in its leaves, often used for making tea.

Chapter 5

Lawn Plants

Bluets - *Hedyotis (=Houstonia) caerulea*
Common Chickweed - *Stellaria media*
Common Dandelion - *Taraxacum officinale*
Common Plantain - *Plantago major*
Common Speedwell - *Veronica officinalis*
Corn Speedwell - *Veronica arvensis*
Curled Dock - *Rumex crispus*
English Plantain - *Plantago lanceolata*
Gill-over-the-ground - *Glechoma hederacea*
Indian Tobacco - *Lobelia inflata*
Kidney-leaved Buttercup (or Aborted Buttercup) - *Ranunculus abortivus*
Moneywort - *Lysimachia nummularia*
Pale Plantain - *Plantago rugelii*
Silvery Cinquefoil - *Potentilla argentea*
Slender Speedwell - *Veronica filiformis*
Thyme-leaved Speedwell - *Veronica serpyllifolia*
White Campion - *Lychnis alba*

6

Bog Slogging

THERE IS A certain primordial atmosphere about a bog. I almost expect to see some prehistoric animal emerge from the ooze. When I go bogging, nothing quite like this happens, but in spite of this, I find bogs fascinating places to explore. They are filled with many plants that can adapt to conditions intolerable for most plant species. There are unusual looking plants, like the pitcher plant, sundew and bladderwort. There may be thousands of sundew plants in a single large northern bog. These carnivorous plants all have special devices for catching organisms to supplement their nitrogen. Some bogs have beautiful orchids like the rose pogonia and grass pink.

For many years it was believed that bog plants had many of the same characteristics as desert plants. The term "physiologically dry" was used to explain what seemed to be the inability of bog plants to get sufficient water for their needs.

It was thought that the great absorbing ability of the sphagnum moss that makes up the bog mat and the cool temperature in the bogs made it difficult for plants to absorb water through their roots. Recent studies indicate that "physiological drought" does not exist. Bog plants can obtain sufficient water, but they have trouble getting the proper nutrient concentration.

Four of the most common shrubs found in bogs are leatherleaf, bog laurel, labrador tea and bog rosemary. These are evergreen plants characterized by thick slender leaves that are succulent, waxy and often with hairy undersurfaces. They all have leaves that are inrolled. Being evergreen appears to be an adaptation related to nutrient conservation (especially phosphorus). The hairs on the undersurface of the leaves are believed to be a protection against the cold so that the plants can function later in the fall and earlier in the spring.

Another adaptation is the ability of these plants to tolerate a wide range of temperature between the temperature of the roots and that of the upper leaves. Some of these plants are able to blossom even when their roots are encased in ice.

Bogs are inland, wetland areas that usually develop in depressions where the drainage is poor, where there is no outlet or inlet. Many of them form in lakes that were originally kettle holes. This is why there are many bogs in Canada and northern portions of the United States. Bogs become smaller and less frequent as we approach the margin of the glaciers' former advance.

Bogs are low in the number of plant nutrients they possess. Oxygen is lacking and because decay is slow, plant life is not easily recycled. Where sphagnum is present, the water is very acid, made so by the chemical action of the sphagnum.

A somewhat similar pattern of vegetation is found in most bogs, but there are still many variations. This is particularly true if we compare northern bogs with those found in other parts of the country.

Typically, bogs start out as open water of lakes or ponds, with vegetation in the form of sphagnum moss and sedges. Eventually, a bog mat forms, becoming covered with low-growing evergreen shrubs such as bog laurel, leatherleaf, bog rosemary and labrador tea. Bogs may not have all of these shrubs, but they will have some of them. Most of these shrubs cannot tolerate very wet conditions with the possible exception of leatherleaf. It can form mats over water. The dominant plant in the group is usually leatherleaf. Labrador tea tends to be the dominant shrub when leatherleaf is not part of the mat.

Along the outer rim of the bog mat are plants such as buckbean, which spreads by lengthening its underground rootstalk, and swamp loosestrife, which arches over the water and grows new shoots at the tips of its branches. This helps to fill in the lake. In this part of the lake wild calla may also grow. The open water beyond may contain species of pondweed and water lilies as well as species of bladderwort.

Eventually, the center of the lake may fill in and coniferous trees such as tamarack and black spruce invade from around the edges. Dwarf birch may also be present. These trees may in time fill the basin or depression completely as the bog becomes thick and heavy enough to reach the bottom. At this time, the open bog has become a bog forest.

When a lake becomes completely filled in with sphagnum and its woody associates, it is called a muskeg. This is an Indian word originally meaning the rounded tussocks of sphagnum that form from the pressure created by the filling in of a bog.

Exploring a bog mat can be an exciting experience, but because there may be holes in the bog mat, it is important not to go alone. A group is probably best because slipping into a hole in the bog mat can be a scary experience. My experience has been that it is more frightening than dangerous.

Some bogs may have poison sumac growing around them. One of my first bog experiences was walking through what seemed like miles of poison sumac before we got to the bog. It was early spring and the snow was still on the ground. We were all wearing coats and gloves, so perhaps that was why no one was affected.

We will be taking a trip to two northern bogs that have extended southward into the deciduous forest region of New England. The first bog is only about two miles from where I live, the suburb of a small city.

On the outer edges of the first bog are many typical wetland plants such as cinnamon fern, sweet pepperbush, high bush blueberry and swamp azalea. Some of the plants growing farther back from the edge on somewhat drier land are red maple, white pine, shadbush, meadowsweet, wild sarsparilla and starflower.

Farther out on the sphagnum bog mat is a dense cover of leatherleaf, chokeberry, some small tamarack, pitcher plants, sheep laurel, spatulate-leaved sundew, large cranberry and small cranberry.

In the center of the bog where there is still water, there is horned bladderwort, fragrant white water lily, cotton grass, a sedge, and a number of species of rushes.

The second bog in a nearby town will offer a comparison between one bog and another. This is a black spruce bog that developed in a glacial kettle. New plant species here are rhodora, bog laurel and creeping snowberry. There are no sundews. The two tree species are black spruce and tamarack. There are no orchids present in either bog. This second bog is well filled in. A great many stunted black spruce have literally taken over and are the dominant species.

Bogs in southern New England are smaller and fewer in number because there are not as many spots for bogs to develop.

Some of the plants that are often present in northern bogs may or may not be present in southern bogs. Plants like Labrador tea are at the lower limit of their range and may be missing in some bogs. Bog rosemary is also common in some bogs, but not present in others. Black spruce and tamarack may be more scattered in some southern bogs and in some cases are replaced by Atlantic white cedar, especially near the coast.

There are some environments in southern New England that are bog-like but not true bogs. There may be patches of sphagnum along a lake or swamp loosestrife may be attempting to take over, but nothing more. Some spots may be sites where bogs previously existed but have changed into another type of habitat, such as a red maple swamp filled with sphagnum.

Bogs have many values. One is as an area for scientific study because of its unusual physical conditions and unique plants. Pollen studies of bogs have revealed important information on changes in vegetation and climate over the years. Research is being carried out to see if they have value as a wildlife habitat.

For many years, man has taken advantage of bogs for raising cranberries. The water level of bogs is manipulated to make crops of cranberries more productive. Bogs often are already saturated, but may serve some role in flood abatement.

As a result of satellite mapping, huge reserves of peat, the largest in the world, have been discovered in Canada. This has great potential as a replacement for coal and for use in horticulture because of the absorbent qualities of the type of peat formed from sphagnum moss. However, great care is necessary in understanding the ecological impact that peat mining may have on the environment.

Bogs are fascinating places to visit, not only because of the many unusual and often unique plants found there, but because of the experience itself. You must not be timid about sinking up to your ankles, and sometimes deeper, in mud or water. Heat and humidity accompany you in your exploration; sometimes pesky insects. Despite these drawbacks, I feel sure that a visit will be well worth your while, as I have found it to be. If it is possible, approach the bog by canoe or boardwalk to protect the bog flora. Trampling by humans can cause serious damage.

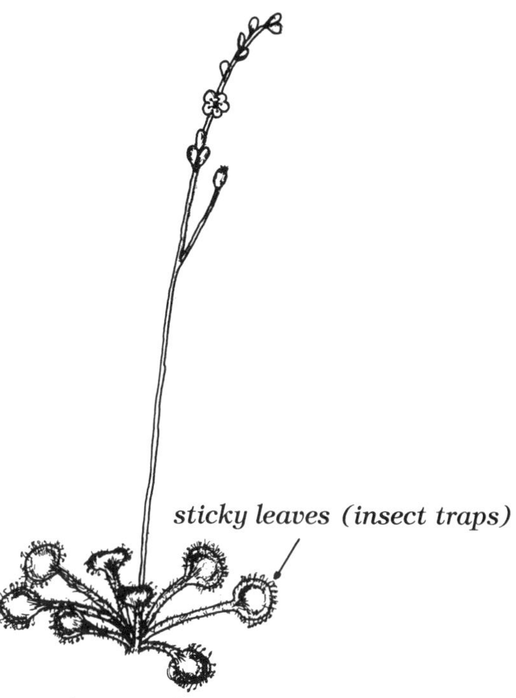

Round-leaved Sundew

(Drosera rotundifolia) Size: 4-9″
Sundew family (flowering stalks)

Round-leaved sundew is one of many species that are widely distributed. The different species can be distinguished by differences in the shape of their leaves. All are plants of swamps and bogs. Two other species found in the Northeast are the spatulate-leaved sundew *(Drosera intermedia)* and, less commonly, the thread-leaved sundew *(Drosera filiformis)*.

The round-leaved sundew is characterized by 1/4 to 1/2 inch wide round leaves covered on their upper surfaces with many red-tipped, glandular hairs. The glands contain a glistening fluid, resembling a dewdrop. The genus name Drosera comes from the Greek word meaning "dewy." The leaves lie flat on the ground arranged in a rosette. Its flower grows on a wiry stalk that is curled over at first, and later unwinds to reveal tiny white blossoms arranged in a one-sided cluster. The blossoming time is from June to August.

The glandular hairs act as an active trap to catch insects. This trap will react only to something that is protein. Their enzymes are capable not only of digesting insects for the nitrogen they need, but of breaking down cartilage and tooth enamel. The acrid juice of the plant was once used for the treatment of corns and warts.

Labrador Tea

(Ledum groenlandicum)
Heath family

Size: 3-4′

Labrador tea grows as a dominant, low growing shrub in many northern bogs. It is also found farther south, but its presence is more scattered. This evergreen shrub is more common as a shrub in forested spruce bogs.

The twigs of Labrador tea form buds in early June and the following year produce handsome white flowers, appearing as terminal clusters. The flowers range in size from 1/3 to 1/2 inch wide. This is followed by the appearance of fruit in the form of a five-parted capsule growing on an incurved stem.

The stems of Labrador tea are densely hairy and the leaves are narrow and waxy beneath, white when young and brown on the older leaves. The edges of the leaves are inrolled.

Early settlers used the leaves as a substitute for tea. It was referred to as "muskeg tea," also "St. James' tea" and "Hudson Bay tea."

The leaves placed among clothing were believed to be useful in discouraging moths. Bees are strongly drawn to the flowers, but browsing animals tend to avoid the plant, believed to be somewhat poisonous.

Leatherleaf

(Chameadaphne calyculata)
Heath family

Size: 1-3′

Leatherleaf is also known as Cassandra from a character in Greek mythology. Its genus name is derived from two Greek words, *chamoi* meaning "dwarf" and *daphne,* meaning "laurel" - hence, a laurel-like plant that grows close to the ground. The species name *calyculata* means that its flower has an outer calyx.

Leatherleaf is a low, many-branched shrub of bogs and swamps. It grows in bogs at the edge of the mat where it comes in contact with open waters. This shrub develops from a very compact interwoven root system. This root system tolerates water and flooding, and often grows suspended several feet thick above the water that lies beneath.

The white, bell-like flowers of leatherleaf blossom very early in the spring, sometimes in March, before the leaves appear. They develop in the axil of the leaves as a nodding raceme. The flowers start developing during mid to late summer of the previous season and although about to flower, it winters over without blossoming.

Leatherleaf has leaves that are evergreen, alternate and leathery, as the common name implies. Their undersides are somewhat yellow, scaly and inrolled. The leaves are eaten at times as browse by the white-tailed deer, especially in northern states like Wisconsin.

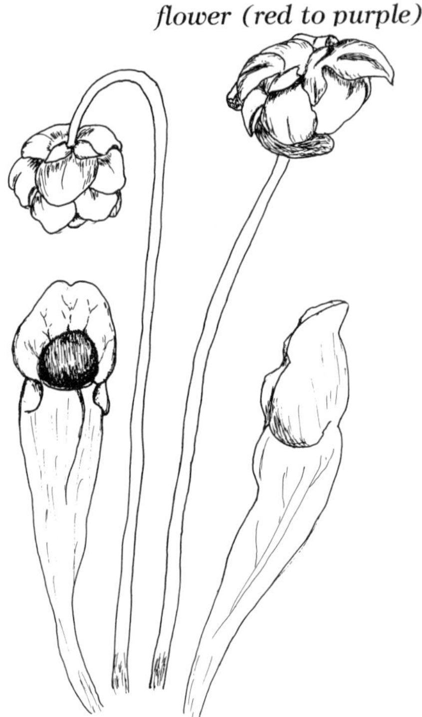

Pitcher Plant

(Sarracenia purpurea)
Pitcher Plant family

Size: 8-24″

This common carnivorous bog plant consists of several hollow pitcher-shaped leaves that curve upward from a buoyant basal platform, usually lying buried among the sphagnum moss of a bog.

Each green to reddish leaf has a wing or seam along one side and ends in a recurved hood at the top. This acts as a landing platform for insects. Near the opening there is nectar on the red veining. This surface is covered with stiff hairs, pointing downward. Beyond this lies a smooth waxy surface. Flies attracted to the plant fall into the water that has collected at the bottom and are usually unable to escape. They are digested by enzymes and bacterial action to supply the plant with the nitrogen it needs.

The single, usually red-purple flower appears on a leafless stalk some time between May and August. An unusual feature of the flower is its style which expands at the top to become umbrella-shaped with five lobes. These lobes hang downward and contain the stigmas.

The pitcher plant was named for a Dr. Sarrazin who was taught by the Indians how to make a medicine from the roots for the treatment of small pox.

Bog Laurel

(Kalmia polifolia)
Heath family

Size: 1-2″

The noted taxonomist, Linnaeus, named the genus *Kalmia* after his close friend and pupil Peter Kalm, a Swedish botanist and traveler. this genus includes, in addition to bog laurel, the more familiar mountain laurel *(Kalmia latifolia)* and sheep laurel *(Kalmia augustifolia).*

Bog laurel is usually found in bogs or on mountain tops, growing as either an erect or trailing shrub. The magenta flowers of bog laurel appear from May to July. They are from 1/2 to 3/4 inch wide and grow at the tips of the stems in rather sparse clusters. Sheep laurel has a similar flower but it blooms along the leaf axil.

The flower of bog laurel, like that of mountain laurel and sheep laurel, has a remarkable adaptation for cross-pollination. The fused petals are wheel-like with 10 pits for holding each of its stamens. The pollen is scattered over the body of a visiting bee when the stamens are touched.

The leaves of bog laurel are narrow and evergreen. Their white undersides are rolled in to conserve moisture. Its twigs are two-edged.

Grazing, domesticated animals have been poisoned by eating the leaves of bog laurel, but this is usually not a problem because it usually grows on inaccessible sites. Deer do not appear to be affected by browsing its leaves.

Chapter 6

Plants Found in and Near Bogs

Atlantic White Cedar - *Chamaecyparis thyoides*
Black Spruce - *Picea mariana*
Bog Laurel - *Kalmia polifolia*
Bog Rosemary - *Andromeda glaucophylla*
Buckbean - *Menyanthes trifoliata*
Chokeberry - *Aronia* sp.
Cinnamon Fern - *Osmunda cinnamomea*
Cottongrass - *Eriophorum virginicum*
Creeping Snowberry - *Gaultheria hispidula*
Dwarf Birch - *Betula pumila*
Fragrant Water Lily - *Nymphaea odorata*
Grass Pink - *Calopogon pulchellus*
Highbush Blueberry - *Vaccinium corymbosum*
Horned Bladderwort - *Utricularia cornuta*
Labrador Tea - *Ledum groenlandicum*
Large Cranberry - *Vaccinium macrocarpon*
Leatherleaf - *Chamaedaphne calyculata*
Meadowsweet - *Spiraea latifolia*
Pitcher Plant - *Sarracenia purpurea*
Poison Sumac - *Rhus vernix*
Pondweed - *Potamogeton* sp.
Red Maple - *Acer rubrum*
Rhodora - *Rhododendron canadense*
Rose Pogonia - *Pogonia ophioglossoides*
Round-leaved Sundew - *Drosera rotundifolia*
Shadbush - *Amelanchier* sp.
Small Cranberry - *Vaccinium oxycoccus*
Spatulate-leaved Sundew - *Drosera intermedia*
Sphagnum - *Sphagnum* sp.
Spicebush - *Lindera benzoin*
Starflower - *Trientalis borealis*
Swamp Azalea - *Rhododendron viscosum*
Swamp Loosestrife - *Decodon verticillatus*
Sweet Pepperbush - *Clethra alnifolia*
Tamarack - *Larix laricina*
Wild Calla - *Calla palustris*
Wild Sarsaparilla - *Aralia nudicaulis*

7

The Plants on Apple Road

EVERY SUBURBAN COMMUNITY has its interesting islands and patches of green where you can study the native vegetation near at hand. This is the case on Apple Road near my home, a short connector road between two streets. It is probably not more than 300 yards in length, but here I have enjoyed many hours of botanizing.

I have asked myself what it is about this road that makes it so intriguing botanically. Perhaps it is the variety of habitats found there and the very recent history of the area. Up to the 1960's the surrounding land was an apple orchard. There is telltale evidence of this from the occasional apple or pear tree that has been left by the developers. This lends a certain charm each spring as the apple and pear tree leaves emerge from their winter buds and their pink-tinged blossoms unfold. These trees are practically identical in the way that they produce spur shoots from which the flowers, and eventually the fruit, form. Only in the nature of the fruit is one plant different from the other.

Since new houses have been built, the land has been managed in a variety of ways. Some lawns are covered with ornamental trees and shrubs. On one lot, a wooded area has been left as part of the landscaping. Another was planted with weeping willows, these were later replaced by spruce trees to hold the soil around a pond. An unmanaged lot that has been left to return to nature is the most interesting botanically. This spot and a patch of staghorn sumac are two prime examples of nature's way of taking over areas when they are left untouched by man. These areas are first occupied by sun-loving plants. They may later be replaced by more shade-loving plants.

The gutters along the road and the edges of lawns could easily go unobserved. Here grow some of the smaller but none-the-less intriguing plants. With the help of a hand lens, these inconspicuous plants reveal many fascinating details.

I take many walks along this road, so although the plants remain the same, they are often in different stages of development. As I start my walk, the first plant I notice is a common roadside plant, the white campion, or evening lychnis. It has five, deeply-notched petals, that may, at times, appear to be more. The blossom usually opens in the evening and is closed the next day.

As I continue along the road the small white-petaled flowers of the multiflora rose catch my eye. Heavy curved thorns and an arching growth pattern characterize this shrub. The hips of this plant contain a lot of vitamin C and are considered a wild edible.

Forming a considerable stand above the multiflora rose is the staghorn sumac with its characteristic red fruit and its velvet-like stems, these are only present on the current year's growth. Sumacs are sun-loving and early pioneers in places that are left undisturbed by man. They quickly move in and take over.

Beyond the clump of sumac, branches of an eastern cottonwood extend slightly out into the street. This tree is commonly found on wet bottomland, but it occasionally shows up as a street tree. Its conspicuous sticky winter buds and pendant catkins are dominant features, and when the fruit of the tree releases its seeds, there is a burst of cotton-like material, the reason for its common name.

In a yard a short distance farther along, there is a pear tree growing. Beautiful in fruit and flower, it is a reminder of the original orchard that included quite a section of this area.

The end of the road is near, and the dominant plant life now is the many spruce trees that have been planted along one side of a pond. It is along this stretch that my attention turns to some of the grasses that grow along the road near the curb. To understand grasses you need to learn a new set of terms to describe their parts and a hand lens to observe the details, particularly the flower. During the fall, crabgrass, the fall panic grass and the purple top are all in flower. They are usually discounted as weeds, but if you are looking for beauty of detail and form, they are well worth examining more closely.

It is now time to double back and see what there is on the other side of the road. Along the curb, extending slightly over the edge, I notice the silvery cinquefoil, with its five-petaled yellow flower and its white woolly leaves. This is a common place to find it growing.

The vacant lot that has been allowed to go back to nature is on this side of the road. It is amazing to see the variety of plant life in such a small area. The predominant plant here is the quaking aspen, commonly referred to as poplar or popple. It is a common pioneer plant on what is sometimes called post-agricultural land.

Parts of this lot seem to have rather poor soil as seems indicated by great masses of the beautiful pink-earth lichen. Here and there a patch of British soldiers lichen appears along with a small type of haircap moss with its transparent or whitish-looking awn, a bristle-like extension of the leaves.

Clumps of sweetfern also present suggest the same type of environment. This is not a true fern, but a shrub with fern-like leaves.

One of the most appealing of the plants that take over land left unoccupied and undeveloped is the black locust. It was originally a tree of the southern Appalachians, but has since become naturalized and is not considered a pioneer plant in New England. Black locust is a legume having nitrogen-fixing bacteria on its roots, helping it to enrich the soil.

The attractive wild indigo, another legume, grows near the locust. This plant bears yellow blossoms in June, and the leaves turn black in the fall. The fruits that form are small pods that rattle. This plant is sometimes called "rattleweed" for this reason. Wild indigo was once used as an inferior substitute for the true indigo dye.

The unused lot seems to have its share of legumes. Besides black locust and wild indigo, there is bush clover and a small-leaved type of tick trefoil, containing small pink flowers.

Leaving this spot and moving on, I notice a patch of delicate white star-like flowers growing along the side of the road near the edge of the curb. This is carpetweed, sometimes referred to in uncomplimentary terms as the Uriah Heep among weeds. I examine its five star-like white sepals with my lens and find it hard to understand how it could be so called.

We now tend to think more in terms of the roles that plants play in their environment rather then emphasizing knowing their names. However, I can't help thinking of the following quotation by Richard Jeffries, when I look back on the pleasure I have experienced exploring this short road over the years. He wrote, "The first conscious thought about wildflowers was to find out their names . . . the first conscious pleasure . . . and then I began to see so many that I had not previously noticed. Once you wish to identify them, there is nothing that escapes, down to the little white chickweed of the patch and the moss of the wall."

Finding a road such as the one I have described in **your** neighborhood could open for you a whole new world for exploration and enjoyment.

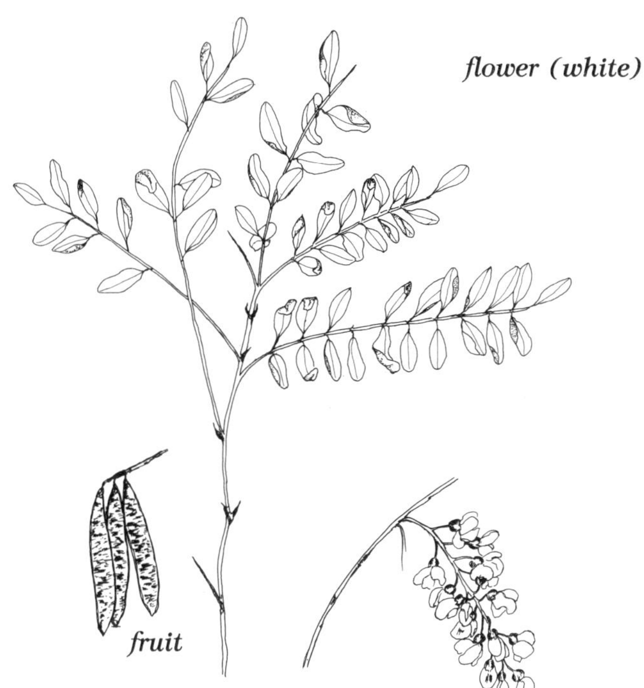

Black Locust

(Robinia pseudoacacia)
Pea family Size: 70-80′

Black locust is an interesting tree at all seasons. Its white pea-like blossoms appear from May to June. When they develop, they are very distinct in their drooping spike-like clusters and are very fragrant.

The leaves are compound, consisting of 7-19 rounded leaflets. Its fruit is a dark brown, two to four inch pod, that matures in early fall and often remains on the tree all winter.

The stem is somewhat zigzag and has paired spines at each node, this is where leaves arise. The winter buds appear to be lacking, but they are hidden beneath the leaf scars. In the spring, the white, hairy buds become enlarged and push through the leaf scars.

The bark is also very distinctive, being very deeply furrowed. This encases a very hard and durable wood having many uses if it were not for the locust borer beetle that damages the wood. Historically, black locust was used for corner posts for early colonial houses and for making wooden nails that were used for shipbuilding. Today, it is largely used for railroad ties, fence posts and fuel. Its value for wildlife is limited, but mourning doves and cottontail rabbits sometimes eat the seeds, while white-tailed deer may browse on the foliage.

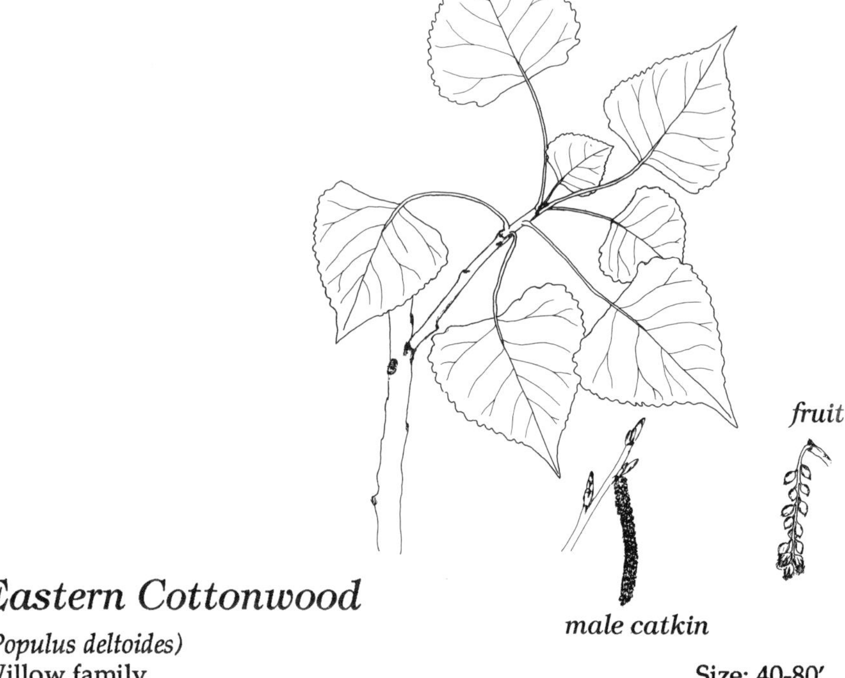

Eastern Cottonwood

(Populus deltoides)
Willow family

Size: 40-80′

The common name for this tree is quite obvious if this is observed in the spring. It refers to the abundance of cottony seeds released from a long string of oval capsules. These hang downward and usually burst open sometime between April and June. The seeds are in such quantity that they often cover the ground and resemble snow.

Because the oval capsules are arranged like a string of beads, cottonwood is also called "necklace poplar." The small inconspicuous male and female flowers blossom from February to May before the leaves expand. They are in the form of catkins and appear on separate trees.

The leaves are triangular-shaped with sharply pointed tips. The leaf margin is toothed while the leaf stalks are flattened. Autumn brings an attractive color to the leaves as they vary from yellow to gold to orange. Cottonwood has large sticky winter buds and yellowish-green bark that changes with age to a fissured gray.

This species of poplar is fast-growing but short-lived. The heartwood begins to decay at about 75 years. This allows woodpeckers to excavate the soft wood for their nests, while yellow bellied sapsuckers use the tree for food and cover, as well as a nesting cavity. A few birds such as the purple finch and evening grosbeak feed on the buds.

White Campion

(Lychnis alba)
Pink family

Size: 1-3′

The generic name of this plant was derived from the Greek word *lychnos* meaning "lamp." The name was probably inspired by one of the brightly colored species such as scarlet lychnis or maltese cross, a European species.

Just under the petals, the five sepals unite to form another lamp-shaped feature, the calyx. The flower of evening lychnis is white, but may occasionally be pink. The flower blossoms at night and is closed the next day. It is very fragrant and attracts moths that pollinate the flower. The blossoming time is usually from July to October.

The female and male reproductive parts are on separate plants. The female flower has a large inflated calyx and five curved styles extending out from the center. The male has a slender calyx and 10 stamens. Evening lychnis is sometimes confused with night-flowering catchfly and bladder campion, but both of these plants have only three styles and the sexes are both in the same flower.

There are some interesting superstitions connected with this plant. One of these is that your mother will die if you pick this plant. Another superstition is that you will be struck by lightening if you pick it.

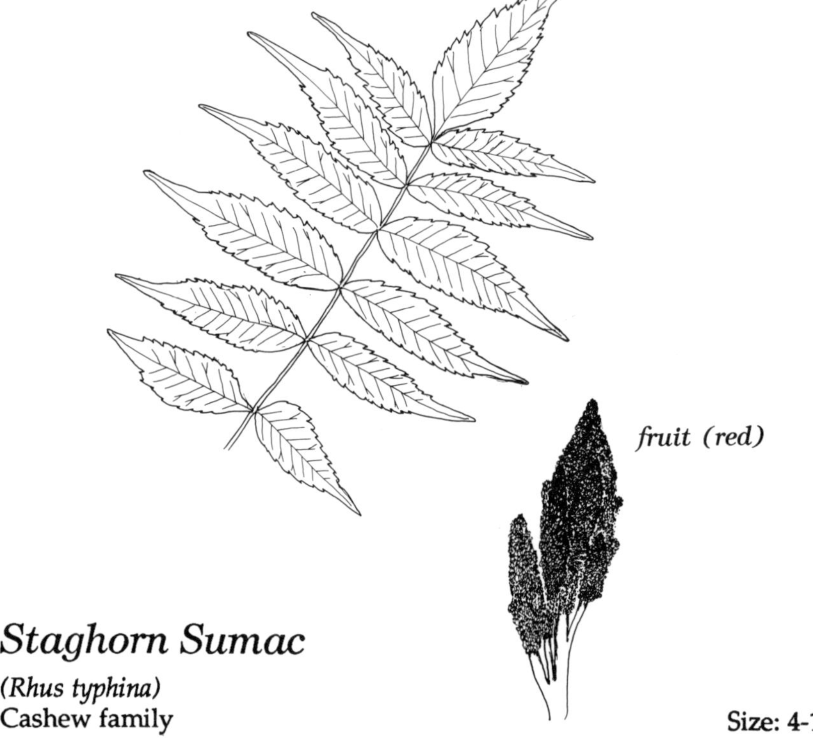

Staghorn Sumac

(Rhus typhina)
Cashew family

Size: 4-15′

Staghorn sumac is one of the largest of the sumacs, appearing as a small tree or large shrub. It is often found growing in clumps, in old fields, pastures and clearings. It has a straggling, awkward growth habit, but in spite of this, it puts on a spectacular display of color in the autumn with its compound leaves that may be orange, red or even purple. Because of its colorful fall foliage and attractive red fruit, it is sometimes used as an ornamental to attract wildlife.

The name "staghorn" refers to the similarity between the branching new growth covered with dense hairs and the antlers of a deer when they are "in velvet."

Its yellow-green flowers are inconspicuous and form in a terminal blossom from May to July. The bright red fruit is small, hairy, one-seeded and points upward. It appears in closely packed clusters from August to September and remain on the tree until spring.

Poison sumac, a species with which it is sometimes confused, has white berries that hang downward. This species of sumac is confined largely to swampy areas.

Many birds feed on the fruit of staghorn sumac, including mocking-birds, cardinals and catbirds.

Indians smoked the dried sumac leaves or mixed them with tobacco and other leaves.

Sweetfern

(Comptonia peregrina)
Wax-Myrtle family

Size: to 5′

The sights and scents that we experience in our environment help our senses to come alive. Sweetfern emits a fragrant aroma that comes from tiny resinous glands that are present on the leaf. The odor is strongest when the plant is crushed, but may be evident after a rainfall, or when there is dew on the ground.

The common name "sweetfern" is misleading because it is not really a fern but a small shrub. The leaves of this plant have rounded lobes similar to some ferns, hence the confusion.

Unlike ferns, sweetfern is a flowering plant that reproduces by means of flowers called catkins. The male and female catkins may be on the same or on different plants. They appear from April to June. The fruit is burr-like with a small shiny nut inside.

Sweetfern grows as a pioneer plant on open, barren, sandy soil such as pastures and old fields. It was once used as a medicine in a variety of ways ranging from treatment for a toothache to rheumatism and childbirth.

Birds such as ruffed grouse eat the buds and catkins of sweetfern and the fruit is eaten by mourning doves and flickers. Deer browse the foliage.

Chapter 7

The Plants on Apple Road

Apple - *Malus* sp.
Black Locust - *Robinia pseudoacacia*
British Soldier Lichen - *Cladonia cristatella*
Carpetweed - *Mollugo verticillata*
Crabgrass - *Digitaria sanguinalis*
Eastern Cottonwood - *Populus deltoides*
Fall Panic Grass - *Panicum dichotomiflorum*
Juniper Haircap Moss - *Polytrichum juniperinum*
Multiflora Rose - *Rosa multiflora*
Pear - *Pyrus* sp.
Pink Earth Lichen - *Baeomyces roseus*
Purpletop - *Triodia flava*
Quaking Aspen - *Populus deltoides*
Silvery Cinquefoil - *Potentilla argentea*
Spruce - *Picea* sp.
Staghorn Sumac - *Rhus typhina*
Sweetfern - *Comptonia peregrina*
Tick Trefoil - *Desmodium* sp.
White Campion - *Lychnis alba*
Wild Indigo - *Baptisia tinctoria*

8

Flowers of Waste Places

IT IS SURPRISING the number of beautiful plants that can be found in unexpected places that are a part of our everyday life. I have found beauty in vegetation along railroad tracks, in small islands of green in cities, in sidewalk cracks, on the edges of lawns near the sidewalk, in the strip of vegetation lying between a sidewalk and the road and in vacant lots. We tend to walk past flowers growing in such places day after day without realizing that they are there, except in a general way. This is especially true for some of the smaller flowers needing at least a hand lens to be appreciated.

I was once riding by a busy shopping plaza when I noticed a number of blue flowers growing in a rather barren area near the highway. When I stopped to look at them, I was thrilled to discover a very beautiful but somewhat bizarre flower, the vipers' bugloss, a very bristly plant with bluish-purple flowers and long projecting red stamens. At a later date, I was to find a large concentration of this exotic plant near another shopping area. Unfortunately, it has since disappeared, as a restaurant now occupies this site.

Around shopping plazas, and at the edge of an industrial site, I often find an unusual looking maple tree . . . the box elder with its un-maple-like leaves. It may even hug the edges or sides of buildings. Another tree that grows under similar conditions is the fast-growing "tree-of-heaven" that some people mistake for sumac.

One day when I went to get a haircut, I noticed a horse nettle growing beside the barber shop in a small three-by-four foot space. This small plot was obviously furnishing a suitable habitat for this plant. Here, it is a thing of beauty; elsewhere it might be a nuisance.

The blossom of this plant is usually white-green, but it may be blue or lavender. The plant is a member of the same family as the potato, tomato and eggplant, and it contains a poisonous chemical, solanine. However, herb doctors used to prepare its orange berries, containing only small amounts of the poison, by drying them. The preparation was then used as a sedative and anti-spasmodic. The berry was also once popular in the treatment of epilepsy.

Nearly everyone is familiar with the common dandelion, but probably a great many people are not aware of the fall dandelion, a different species, that blooms from June to November. The flower stalk of this species does not have milky juice like the common dandelion, and the underside of the outer ray flower is usually reddish. The flower head itself is not as full as that of the common dandelion. Often fall dandelions appear as a few isolated specimens that grow on lawns and in the space between the sidewalk and the road.

Another plant that commonly grows along this strip is shepherd's purse. It also grows around buildings and on lawns. The species name, *bursa-pastoris* is Latin for "shepherd's purse" and was given because of the resemblance of the flat seeds to a purse. The seeds can be used to add flavor to soups and salads. The young leaves are also edible. When they are boiled, they reportedly taste like cabbage.

The area around railroad tracks offers a wealth of opportunity for the plant lover to find a wide variety of species. Many alien plants become established here, and along roadsides because they do not find competition from other plants. Aliens are plants not native to an area, but become established, and the seeds being brought here in a variety of interesting ways.

One of my most exciting discoveries while botanizing along railroad tracks was the four-o'clock *(Mirabilis nyctaginea).* The specimens I found were red, but they may be found in gardens as pink, purple and yellow varieties as well. The generic name in Latin means "remarkable" or "wonderful." The species name is Greek for "night-blooming." Thus, a wonderful plant that opens late in the afternoon - at about four o'clock.

A number of different plant species manage to survive even in the cracks of a sidewalk. They live in spite of little or no soil, lack of moisture and vastly fluctuating temperatures. These include plants such as the smartweeds, prostrate knotweed, dandelion and some species of plantain.

The richest of the waste places to botanize is probably the vacant lot. Often it will yield a remarkable array of interesting specimens. Here I have made such discoveries as ladies' tresses (an orchid), wild sensitive plant, rattlebox, whorled milkwort and sand jointweed.

The vacant lot I frequently visit is a field that has been unused for many years and there has never been any construction on this property. Here, I find nodding ladies' tresses growing at the edge nearest the road. This is a small orchid with flowers that arch downward, hence the specific name "nodding." The Greek generic name means "spiral-flowered," referring to the twisted flower stalk of this plant.

Many specimens of wild sensitive plant grow nearby. This is a tiny short-stalked yellow flower that is a legume related to the common garden pea. Its leaves have the ability to fold when subjected to mechanical stimulation, thus, the reason for the term "sensitive."

Rattlebox is well named. This refers to the rattling sound made by the ripe seeds of this legume when its hard pod is disturbed in some way. Its generic name *Crotalaria* is a Greek word meaning "rattle," the same as its common name. The species name *sagittalis* is a Latin word referring to the arrow-shaped stipules. These are small ear-like appendages found at the base of the leaves.

Two small, but exquisite flowers are the whorled milkwort and sand jointweed. Both these plants grow along the edge near the street. The whorled milkwort has short linear leaves, usually in whorls of three to six. Leaves that are in whorls are those consisting of three or more, all appearing at the same level on the stem.

The sand jointweed, as its name implies, grows in sandy soil. It has very small pink or white blossoms and the stem is wiry and jointed.

One of the more typical flowers found on this lot is the common St. Johnswort with its yellow flowers resembling small powder puffs. This effect is created by the presence of its many stamens.

Some of the common aliens found on vacant city lots are white clover, red clover, mugwort, ox-eye daisy, St. Johnswort, white campion, burdock, bouncing Bet, yellow and white sweet clover, Japanese knotweed, chicory, curled dock, lady's-thumb, lamb's-quarters, Mexican tea, milkweed, motherwort, Queen Anne's lace, mullein and many others that you can add to the list. Each site has its own particular group of colorful and interesting plants.

Red Clover

(Trifolium pratense)
Pea family

Size: 6-24″

We generally think of red clover as a plant of grassy areas in a rural setting. This is true, but we also find clover quite commonly in waste areas such as vacant lots. It grows well here because it has nitrogen-fixing bacteria on its roots, helping it to enrich the soil.

Clover was once known as *clava,* the Latin word for "club." It was thought that the clover leaf resembled the three-knobbed club of Hercules. Over the years *clava* became corrupted to clover.

All clovers must depend on bees if they are to be fertilized. The reddish flower head may have from 40 to 60 florets with each floret resembling a miniature sweet pea flower. The stamens and pistil are enclosed in two fused petals. These must be opened to allow pollen to escape.

Bumble bees, because of their large size, are the best pollinators of red clover. Domesticated honey bees are less effective and more often seek the small white clover to produce their honey.

Early farmers used red clover in making a salve that was useful for skin ailments. Today, we use the leaves and flower heads for salads and as a potherb.

Tree of Heaven

(Ailanthus altissima)
Quassia family

Size: 80-100′

Betty Smith's popularized *Ailanthus* in her book, ***"A Tree Grows in Brooklyn."*** It was source of inspiration for a young girl who admired its ability to adapt to a hostile environment.

Ailanthus can grow nearly anywhere, even through cracks in a sidewalk. It is sometimes confused with sumac, but most sumacs have toothed leaves and fewer leaflets. In *Ailanthus,* the compound leaves have teeth only at the base of each leaflet.

The greenish yellow flowers are inconspicuous and the sexes are usually on separate trees. Blossoming time is from June to July followed by the formation of clusters of propeller-like seeds. Some seeds may persist on the tree during the winter. This, along with its large heart-shaped leaf scars, make the tree easy to recognize, even during the winter months.

Ailanthus was introduced into this country as an ornamental in 1874, but it soon became a pest because of its rapid growth (up to nine feet per year) and its many root suckers.

In spite of these drawbacks, it is a hardy tree that can tolerate the salt, high temperature and dryness of an urban environment, and it possesses an unusually strong root system. These are all characteristics that horticulturists are striving to develop in other urban trees such as sycamores, maples and ashes.

Common St. Johnswort

(Hypericum perforatum)
St. Johnswort family

Size: 1-2½′

Few plants have as many superstitions and virtues surrounding them as the common St. Johnswort. Often associated with St. John's Day, June 24th, one belief was that if bits of the plant were hung over the doors and windows in June, this would scare away witches and evil spirits.

An alien from Europe, it now grows in abundance along roadsides, in fields and in waste places such as vacant lots.

Common St. Johnswort is a plant with narrow, elliptical leaves and a branching growth habit. The flowers appear from June to September, several in a terminal cluster. They are yellow and five-petaled with many stamens. The flower lacks nectar, so it is visited only by pollen-eating insects.

Small oil glands, that may appear as black or translucent, if held up to the light, are found on both the leaves and petals. If white-skinned animals such as domestic cattle eat St. Johnswort, their skin may become very irritated. The poison that causes this problem, is a pigment that is found in the tiny oil glands of the plant. This adverse reaction occurs only when the animal is exposed to excessive sunlight.

It was once believed that the oil from these same glands could be used in cuts or wounds that pierced the skin.

Butter-and-Eggs

(Linaria vulgaris)
Figwort family

Size: 1-3′

Butter-and-eggs resembles a snapdragon, but is much smaller. It is in the same family with the common mullein, foxglove and monkey flower. Before it flowers, it bears a close resemblance to cypress spurge (*Euphorbia cyparissias)* because it grows to a similar height and the leaves look much alike. However, cypress spurge has a milky juice that may be irritating to the skin.

The flowers of butter-and-eggs bloom from late June through October. It usually grows on sites that are not very desirable for other plants. Introduced from Europe, it is a common plant of roadsides and waste places.

The name "butter-and-eggs" stems from the flower's two shades of yellow that resemble butter and egg yolk. The genus name *"Linaria"* originated because its leaves resemble those of flax.

This plant has a very interesting adaptation to ensure that pollen is picked up by the right insect. Its flowers are two-lipped, and can only be opened by certain insects that are heavy enough to pry open the flower. These include bumblebees, honeybees, other wild bees and an occasional butterfly if it is large enough. The insects enter the flower looking for nectar and in the process they become covered with pollen.

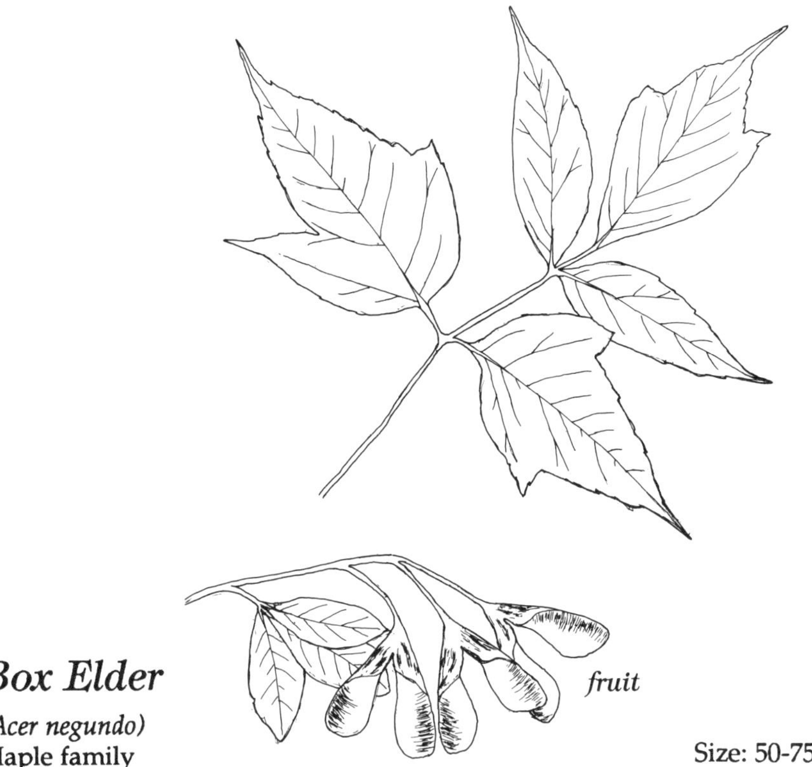

Box Elder

(Acer negundo)
Maple family

Size: 50-75′

Who would guess that this tree is a species of maple? Its compound leaves, when five or more leaflets are present, resemble those of ash or elderberry and when only three leaflets are present, it looks like poison ivy. The one maple-like characteristic is its winged seeds that resemble ice tongs.

Its common name, box elder, comes from the resemblance of the leaves to elderberry (*Sambucus)* and its whitish wood is like that of the English box tree *(Buxus sempervirens).* Another of its common names is Manitoba maple.

Box elder prefers a moist environment such as a flood plain, but it is very adaptable, appearing along roadsides and in waste places.

The tiny male and female flowers appear on separate plants from March to mid-May. They are in dense clusters, each flower hanging from a thin stalk. The male flowers are red and the female flowers are green.

Another attractive feature of the box elder is the appearance of its winter twigs varying in color from green to purple.

Box elder has little value as wood, being used only for crates, boxes and pulpwood. Its seeds are eaten by mice, squirrels and a few birds such as the evening grosbeak and purple finch. Deer browse the young twigs and leaves.

The sap was once used for making maple sugar and maple syrup.

Chapter 8

Plants of Waste Places

Bouncing Bet - *Saponaria officinalis*
Box Elder - *Acer negundo*
Burdock - *Arctium* sp.
Butter-and-Eggs - *Linaria vulgaris*
Chicory - *Cichorium intybus*
Common Dandelion - *Taraxacum officinale*
Common Milkweed - *Asclepias syriaca*
Common Mullein - *Verbascum thapsus*
Common St. Johnswort - *Hypericum perforatum*
Curled Dock - *Rumex crispus*
Fall Dandelion - *Leontodon autumnalis*
Four-o-clock - *Mirabilis nyctaginea*
Horse Nettle - *Solanum carolinense*
Japanese Knotweed - *Polygonum cuspidatum*
Lady's-Thumb - *Polygonum persicaria*
Ladies'-tresses - *Spiranthes cernua*
Lamb's-quarters - *Chenopodium album*
Mexican Tea - *Chenopodium ambrosioides*
Motherwort - *Leonurus cardiaca*
Mugwort - *Artemisia vulgaris*
Ox-eye Daisy - *Leucanthemum vulgare*
Prostrate Knotweed - *Polygonum aviculare*
Queen Anne's Lace - *Daucus carota*
Rattlebox - *Crotalaria sagittalis*
Red Clover - *Trifolium pratense*
Sand Jointweed - *Polygonella articulata*
Shepherd's Purse - *Capsella bursa-pastoris*
Tree-of-heaven - *Ailanthus altissima*
Viper's Bugloss - *Echium vulgare*
White Campion - *Lychnis alba*
White Clover - *Trifolium repens*
White Sweet Clover - *Melilotus alba*
Whorled Milkwort - *Polygala verticillata*
Wild Sensitive Plant - *Cassia nictitans*
Yellow Sweet Clover - *Melilotus officinalis*

9

A Walk Along a Country Road

THERE IS A certain roadside I have visited on many occasions that is especially appealing because of the wide variety of botanical species and differing habitats to be found there.

This particular road dips and curves tending to make you wonder what is around the next bend. At frequent intervals, a new type of habitat, such as a field or swamp, presents itself as it adjoins the road. There is a farm that is being worked and a field that is separated from the road by a wooden fence. Farther along, a number of estates are being managed in various ways. Lines of locust trees and occasional apple and cherry trees offer beauty, particularly when in flower. A solid stand of hemlocks in a valley is a point of interest.

An outcropping of granite restricts the growth of vegetation on the side of the road in one spot. Here, at a glance, we can see the beginning stages in the evolution of plant life because of the primitive lichens and mosses found in and around the bare rock.

Almost directly opposite is the farmer's field mentioned earlier. Here the field illustrates vegetation development . . . how growth of plants develops after having been interfered with by man because of mowing, as opposed to the situation on bare rock where growth has occurred without any interference by man.

The apple trees and cherry trees tell us of past agricultural activity in the area. Cherries represent pioneer shrubs or trees that grow up along roadsides after cutting has been done. The apples were probably planted. The locust trees are another pioneer species, indicating cutting or clearing.

The roadside itself throughout its length of about two miles, shows evidence of a different kind of management . . . constant mowing. This sometimes makes it necessary to look farther afield, to what can be seen on the other side of the fence. The environment here varies from swamp to woods to fields.

A roadside offers botanical interest of one kind or another throughout the year. Each season brings a focus on different groups of plants. During the summer months, it's the wide variety of alien plants that are on stage. We will be taking a closer look at a few of these. They are often found growing along the road because they do not have the same competition there as on other sites.

We begin our walk near the entrance to the road. A patch of odd-looking leaves catches the eye. They are low growing and resemble a horse's hoof. This characteristic provides its common name, coltsfoot. In mid-summer, only leaves are present; in early spring there is an absence of leaves with only its dandelion-like flower being in evidence. Coltsfoot, like the dandelion, is one of the few members of the composite family to blossom in the spring. Its genus name *Tussilago* implies a common use of coltsfoot . . . *tussiss* in Latin means "cough." Many parts of the plant were used for the treatment of colds and other respiratory ailments.

Across the road, a split rail fence separates the roadside from a swampy area. There are many large elephant-ear leaves of the skunk cabbage. The appearance of the plant has changed drastically from its spring guise. Pick a few leaves of a nearby spicebush and notice the lemon-like aroma. Poking through the fence, there is an occasional Jack-in-the-pulpit. Some are showing their green, corncob-like fruit. Later in the summer, or early fall the fruit will turn to a brilliant red.

Nearer the roadside we see the delicate blossoms of the enchanter's nightshade growing. Notice the two tiny petals of the flower of this plant and tiny burr-like fruit that is beginning to form. This plant is misnamed because it does not belong to the nightshade family. Its proper classification is in the evening primrose family.

Horse nettle is occasionally found along this stretch of road. It is not a nettle but a member of the nightshade family. So much for common names. Sometimes they are rather confusing, but often they are colorful and descriptive of the plant's use or have an interesting story behind them.

We continue on our way noticing some of the plants that seem to hug the edge of the roadside. Here we find sheep sorrel, smartweeds and the common ragweed. Ragweed, and not goldenrod, is the chief offender for the attacks of hay fever. Its genus name, *Ambrosia,* seems to suggest that someone had a strange sense of humor when it was named. Less common, but present along the roadway, is the giant ragweed, growing much taller.

Let's pause for a few minutes to examine a young black walnut tree growing near the fence. With a jack knife, cut into a small branch to notice the light brown, chambered pith. Butternut, or white walnut, has similar chambered pith, but it is darker brown. Notice the appearance of the end leaflet of the compound leaf of the black walnut. It is quite small compared with the other leaflets, or it may be missing. In butternut, the end leaflet is present and often larger than its other leaflets.

On the other side of the fence, there is a massive stand of tall reed grass, or phragmites. Its presence indicates that the soil is wet. Phragmites is common on spots along the seacoast, but has little wildlife value other than cover. It quickly takes over disturbed areas.

Ahead of us is an open field where we will have a chance to make many observations. On our way we notice the presence of blackberry bushes and a tangle of grapevines. Here we have the opportunity to look for the ripening fruit of the jewelweed, or touch-me-not. We pause just long enough to pop one or two of the seed pods to see how quickly they explode if they are fully ripe. On both sides of the road there are the beautiful pea-like blossoms of the black locusts.

The stretch of roadside beside the open field has an amazing array of plants, including some British soldier lichen. This tiny plant with its colorful red tips can be discovered by looking closely on the upper rail of the split rail fence nearby.

Several alien plants are dominating the vegetation here. Pull up a root of Queen Anne's lace and smell its carrot-like odor. This is one quick way to tell the difference between this plant and yarrow. These two plants are often confused, but when examined side by side, there are noticeable differences. Besides the difference in the arrangement of their white flowers, yarrow has a very aromatic herb-like smell. My mother used to steep the leaves of yarrow to make a tea. The tea, as she made it, caused profuse sweating and was helpful in breaking up a fever.

This is a good spot to look for the male and female flowers of the white campion, or evening lychnis. The sexes are found on separate plants that usually grow near each other.

Along the edge of the road, there is a great deal of chicory growing, usually found side by side with Queen Anne's lace.

Nearby is herd's grass, or timothy, that will reward you on closer examination, if it is in flower. A wind-pollinated plant, the dangling anthers and delicate, feather-like pistils will amaze you with their beauty under the hand lens.

Pick a couple of burdock flowers that grow nearby, and notice how they stick to each other. Nature thought of it first before we had velcro to fasten the cuffs of our jackets.

Ox-eye daisy, black-eyed Susan, bouncing Bet, celandine and occasionally goatsbeard and sneezeweed are other plants that grow in this reasonably small area near the open field.

Only a few roadside plants have been covered on this walk. Over a period of time it is possible to encompass the spectrum of plant life along a roadside. The roadside is a reflection of the surrounding countryside. It includes many species of native trees and shrubs as well as some cultivated ones. Wildflowers, ferns, horsetails, clubmosses, lichens, fungi and algae may colonize the area.

Some of the advantages of roadside botanizing are that plants can be studied without walking great distances, and they are easily accessible without trespassing. This makes the roadside a good place for collecting. Perhaps this may lead to a new interest for you . . . looking for a new species through the countryside.

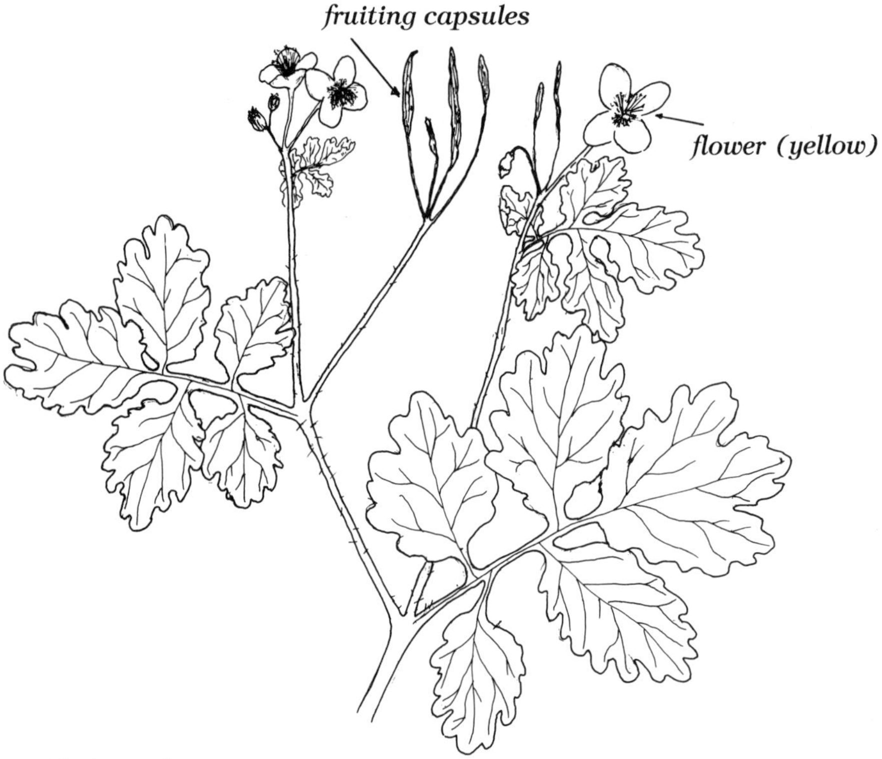

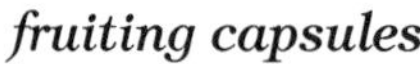

Celandine

(Chelidonium majus)
Poppy family

Size: 1-2′

The small yellow flower of celandine blossoms from April to September. At one time it was believed that flowering began with the arrival of swallows and ended with their departure. The flower appears in clusters on thin yellow stalks, developing from the axils of the leaves. Its two sepals fall off at the time of blossoming giving the impression that they were never present. The plant is somewhat hairy with deeply lobed, compound leaves.

Its generic name *Chelidon* is derived from a Greek word meaning "swallow." Originally celandine was brought to the United States by early New England settlers. They valued it for its medical qualities and planted it in their gardens. It has since escaped and become naturalized.

The leaves, stem and root of celandine produce a yellow juice. An early scientist believed that swallows bathed the eyes of their young in the juice to improve their eyesight. This may have lead to its many uses historically. It was reputed to be useful in removing corns, warts and freckles and in curing sore eyes. There was also the belief that it could be used in the treatment of liver disorders because of the resemblance of its juice to bile.

Ox-Eye Daisy

(Leucanthemum vulgare)
Aster family　　　　Size: 1-3′

The ox-eye daisy is misnamed. It is not a daisy but a chrysanthemum. The word daisy is derived from two Anglo-Saxon words *doegis* and *eage* which together mean "day's eye." This name was originally applied to the common English daisy *(Bellis perennis)*, a true daisy, closing at night and opening when the sun comes up. Like the sun, it was thought of as a day's eye. Others believed the center of the flower resembled an ox's eye.

Extremely widespread, the ox-eye daisy appears commonly in fields and waste places throughout most of the United States. It was brought here originally by the early settlers when they had trouble growing their common English daisy.

The blooming time for the ox-eye daisy is from May to October. It has a composite flower arranged in a head. This consists of many tiny flowers or florets. The white petals are the ray flowers with pistils only. The yellow disk flowers in the center consist of both staminate and pistillate flowers. The leaves are small and narrow, often with many tiny lobes.

Historically, preparations from the leaves of ox-eye daisy were used for burns, bruises and running sores. Another of its common names was bruisewort.

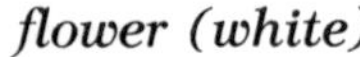

Queen Anne's Lace

(Daucus carota)
Parsley family

Size: 1-3′

One of the most beautiful of the roadside aliens is Queen Anne's lace (*Daucus carota*). The common garden carrot and Queen Anne's lace have both evolved from the same ancestor. Scraping the root or crushing a leaf will release the carrot odor.

The white lacy, flat-topped umbel is composed of many small florets, or individual flowers. The outer florets are larger but sterile. They serve as guides for insects. The center of the umbel often contains a single deep-purple floret. This may help to direct the attention of visiting insects. It is sometimes said that this is where Queen Anne pricked her finger when making lace.

The flower blooms from May to October, and opens according to the amount of moisture in the air. Before the flower blooms and after it has gone to seed, it becomes hollow in the center and resembles a bird's nest. The spiny seeds resemble jewels if you look at them under a hand lens.

The common name can be traced to Queen Anne, wife of James I. It was fashionable at the time for the queen and ladies of the royal circle to decorate their hair with lacy autumn leaves. It is not unlikely that the flowers were used in the same way.

Bouncing Bet

(Saponaria officinalis)
Pink family

Size: 1-2½′

Bouncing Bet was brought to America in the 17th Century and was planted in the gardens of the early settlers. It escaped and has become naturalized.

This perennial blooms from July to October. The flower is usually pale pink, but it is occasionally rose colored. It has a slight fragrance, more noticeably toward night. Because of this fragrance, the petals are sometimes dried and used in potpourri mixtures.

The name, bouncing Bet, alludes to the inflated calyx and scalloped petals of the flower, thought to resemble the rear view of a washer-woman. This association may have been made because another common name for bouncing Bet, is soapwort, and its generic name *sapon* is the Latin for soap.

The leaves, stem and root all contain saponins. These are compounds capable of producing a soap lather when agitated in water. The plant has a long history of being used as a cleaning agent. This started well before the invention of soap.

Bouncing Bet is still used in some parts of the world to restore color and sheen to fragile textiles, old china, precious glass and for the cleaning of oriental rugs. Medicinally, it has been used as an antiseptic for treating wounds, poison ivy rash and other skin disorders.

Chicory

(Cichorium intybus)
Aster family

Size: 1-4′

Chicory is often found growing beside Queen Anne's lace along the very edge of roadsides. The blue, or occasionally white, flower consists only of ray flowers, having a two-parted style. The stamens surround the pistil which, as it develops, grows up through the dark blue anthers surrounding the style. When fully mature, the stigma forms a Y above the stamens. This is best seen with a hand lens.

The flowers are at their brightest in the morning. By noon they often become somewhat ragged. This feature and their blue color is the reason for the common name, ragged sailors. The stalkless flowers appear in clusters of two or three, but only a few open at a time, and they only last a day with new flowers opening the next day. The blooming period is from mid-June to October. The leaves are alternate and quite variable. They may be clasping, toothed, lobed or entire.

Chicory was another of the plants brought to the United States by the early settlers. They cultivated it in their gardens as a source of food. It has long been used as a vegetable and salad plant. Over the years, the ground root of chicory has also proved useful on occasion, either as an ingredient or substitute for coffee.

Medicinally, chicory was once thought to be helpful in the treatment of liver problems, insomnia and for fever reduction.

Chapter 9

Plants of a Country Roadside

Apple - *Malus* sp.
Blackberry - *Rubus* sp.
Black Cherry - *Prunus serotina*
Black-eyed Susan - *Rudbeckia serotina*
Black Locust - *Robinia pseudoacacia*
Black Walnut - *Juglans nigra*
Bouncing Bet - *Saponaria officinalis*
British Soldier Lichen - *Cladonia cristatella*
Burdock - *Arctium* sp.
Celandine - *Chelidonium majus*
Chicory - *Cichorium intybus*
Coltsfoot - *Tussilago farfara*
Common Ragweed - *Ambrosia artemisiifolia*
Enchanter's Nightshade - *Circaea quadrisulcata*
Eastern Hemlock - *Tsuga canadensis*
Giant Ragweed - *Ambrosia trifida*
Goatsbeard - *Tragopogon pratensis*
Grape - *Vitis* sp.
Herd's Grass or Timothy - *Phleum pratense*
Horse Nettle - *Solanum carolinense*
Jack-in-the-Pulpit - *Arisaema triphyllum*
Jewelweed - *Impatiens capensis*
Ox-eye Daisy - *Leucanthemum vulgare*
Purple-headed Sneezeweed - *Helenium nudiflorum*
Queen Anne's Lace - *Daucus carota*
Sheep Sorrel - *Rumex acetosella*
Skunk Cabbage - *Symplocarpus foetidus*
Smartweed - *Polygonum* sp.
Spicebush - *Lindera benzoin*
Tall Reed Grass - *Phragmites australis*
White Campion - *Lychnis alba*
Yarrow - *Achillea millefolium*

10

Plants of a Freshwater Pond Community

THERE IS AN atmosphere of quietness to a freshwater pond. Only the occasional sounds such as the squawk of a green heron, the rattling sound of a belted kingfisher, or the banjo twang of a green frog may break the silence. This is the realm of the painted turtle, the banded water snake and an occasional snapping turtle. Sunfish may rise to the surface once in a while to inspect debris that falls from overhanging branches. Dragonflies and damselflies glide by, sometimes stopping to hover then darting away to look for insect prey. Whirligig beetles spin around hap-hazardly on the surface, like bumper cars at an amusement park.

All these creatures and many more are fundamentally dependent, either directly or indirectly, on the pond plant life community, including both surrounding land and aquatic plants. Some eat the plant material while others feed on the herbivores.

A freshwater pond has its own plant community consisting of many different species. These include both the land plants surrounding the pond and the aquatic plants growing in the water. The land plants may be woody or non-woody, although they thrive in a wetland environment. However, they do not have special adaptations for actually living in the water.

The aquatic plants living in the water are a study in contrast to land plants. Everything about the way they function is the result of special structural adaptations enabling them to survive in a water environment.

Water plants need water, minerals, carbon dioxide and oxygen, the same as land plants do, but because they are surrounded by water, their tissues are modified to adapt to those conditions. For example, they have dissected leaves to increase surface area, air spaces in tissues for storing oxygen, absence or reduced numbers of pores (stomata) for gaseous exchange, and a lack of root hairs.

No two ponds will have exactly the same type of vegetation. It will vary from one pond to another, but in spite of this, there may be various vegetation zones as the pond fills in. These often include the land plants growing along the shore and different zones of aquatic plants growing inward toward the center of the pond.

The aquatic plants of the first zone, closest to the land, are called emergents. These rooted plants have a portion of the stem extending above the water. An example would be the cattails. Farther out, between the shore and the center of the pond, plants like water lilies with roots attached to the bottom, or substrate, but with floating leaves will be found. Pond weeds and other plants in this region have roots attached to the substrate and the whole plant is submerged.

The last group of plants is often found in the deeper waters of a pond. These are free-floating and are not attached to the substrate, but have vegetative parts floating on the surface, like duckweed, and the free submerged plants like bladderworts. The latter plant has its vegetative parts submerged, but its flower and stalk extend above the water.

Let's take a closer look at the way some specific water-loving plants may be distributed in and around a pond. This distribution will, of course, be an over-simplification because the various zones may be more distinct in one pond than another, And there is often much overlap or a zone may be skipped or missing.

The pond that we will visit is rather ideal for illustrating some of the points I have been mentioning. I would like to show you some of the land plants first.

Spicebush does well here. It can be easily identified almost any time of the year by its lemony smell. Crush a leaf, squeeze its olive-shaped fruit or scratch a bit of bark on one of its twigs and the odor is the same.

Growing close by in profusion is sweet pepperbush. An unforgettable plant, it is one of the few shrubs that blossoms late in the summer. On certain days when the sweet pepperbush is in full flower and the atmospheric conditions are just right in the area around a pond, the air will be full of the lilac smell of this fragrant flower. Before it blossoms, which is in late July or early August, it can still be recognized by its last year's spike-like fruit clusters. They resemble tiny peppercorns, but have no odor. I do not know how this shrub got its name, but one possibility is that its flower smells sweet and its fruit looks like peppercorn . . . thus sweet pepperbush.

The swamp azalea is another beautiful wetland plant that grows around the pond margin. It, like the sweet pepperbush, has a very sweet smell. The blossoming time for this species is usually June to July, but sometimes as late as September. This species has a flower tube, quite similar to the pink azalea, or pinxter flower, but has sticky reddish hairs. Look for them with a hand lens. The blossoms are white, although on rare occasions, they are pink-striped to reddish. The species name for this flower *viscosum* refers to the stickiness or viscosity of the buds and flower tube.

Some ponds have poison sumac growing around or near them. This is its natural habitat. There used to be a large clump here, but it has since been removed. It is the only one of our sumacs that is poisonous, producing a dermatitis similar to poison ivy, but probably worse. There is little resemblance between this species and the other sumacs. Its leaves resemble those of white ash somewhat, but there are more leaflets to its compound leaves, and the branches are alternate. White ash has opposite branching. The fruit of poison sumac consists of a cluster of grayish-white berries that hang downward. Our other sumacs have red fruit growing upward.

As we walk around the pond we will notice a considerable amount of skunk cabbage with its large elephant-ear leaves that emerge as it matures. Occasionally we find the fruit of the skunk cabbage. It looks something like a hand grenade. The texture is like a potato and it is ill smelling.

Overhanging the edge of the pond are numerous red maples. When fall comes they, along with other vegetation growing near the edge, will drop their leaves into the water. In doing so, they will aid in the filling in of the pond. Over time, a marsh or a woodland swamp could develop here.

Near the far end of the pond there is an interesting delta to explore. It has been built up gradually over the years. This too, is slowly filling in the pond. Here on the delta, there are many species of plants, including several sedges and rushes. Some of the flowering plants include; ditch stonecrop, blue vervain, arrowhead, water plantain and water purslane. The last three plants grow in the soft mud near the water's edge and in the water itself. On one occasion, I found seedbox, a relative of St. Johnswort. Three species of rushes grow here; the Canada rush, the soft rush and the least spike rush. The only sedge present is the sallow sedge.

At the far end of the pond some years, we find the dramatic looking cardinal flower with its flaming red blossoms.

Another land plant that may form colonial masses along the edges of ponds is purple loosestrife. This is a very aggressive species introduced from Europe It tends to crowd out plants having much more wildlife value. In spite of this, it is a plant of considerable beauty. Its colorful purple flower spikes extend upward, sometimes as much as five feet. Each flower is a study in itself with petals varying from four to six in number and having three different flowers with stamens and pistils of varying length. This is to ensure cross-pollination. Historically, the leaves were used by Indians as a medium for treating chronic diarrhea and when food was scarce, as emergency rations.

A well known rooted emergent plant growing erect along the margins of ponds is the common cattail. A less common species is the narrow-leaved cattail. This has leaves that are less than $^1/_2$ inch wide. The flowering spike in this cattail has the staminate and pistillate flowers separated by a gap. Common cattail has leaves up to one inch wide and the reproductive parts are not separated by a space. If the pond fills in and becomes a marsh, cattails will become one of the most common and conspicuous species.

Another rooted emergent is pickerelweed. This is a relative of the notorious water hyacinth that has clogged the waterways in the south. Pickerelweed, under certain conditions, may also choke up small streams and shallow ponds.

A third rooted emergent is arrow arum. This plant has the same type of flower as Jack-in-the-pulpit and skunk cabbage. They are all members of the arum family characterized by flowers with a spathe and spadix.

Farther into the pond we may find the various types of water lilies, such as fragrant water lily, tuberous water lily, cow lily, and in some ponds, water-shield. Another species that may be present is floating hearts. It has some leaves that resemble water lilies, but it belongs to the gentian family.

Another group of plants with floating leaves is the duckweeds. Some years they appear on ponds to the extent that much of the surface is covered. This may be due to nitrates that come into the pond from runoff.

The bladderworts are a group of delicate plants that may be submerged or floating. They may be anchored, but they lack roots. There are many different species with flowers that are blue, purple, yellow or white. They are all equipped with tiny bladders or traps that catch small organisms in the water. This helps the plant get nourishment.

An unusual plant that I have found in one Connecticut pond was water shamrock. It is a type of fern that has a long stem, enabling its four clover-like leaves to float on the surface.

For those of you who would like to carry your explorations further, many of the submerged aquatic plants can be studied by collecting them in a boat with the use of grapple hooks. Chief among these plants are the many species of pondweed that vary considerably from one species to another.

Underwater parts of plants such as the long fleshy rhizome of the cow lily may require donning rubber goggles and flippers in order to bring this plant material to the surface.

We have been looking at only a limited number of the many plants found in and around a pond. A study of their distinctive adaptations, some of which we have been introduced to already, has a special appeal that I hope will arouse your interest in further exploration.

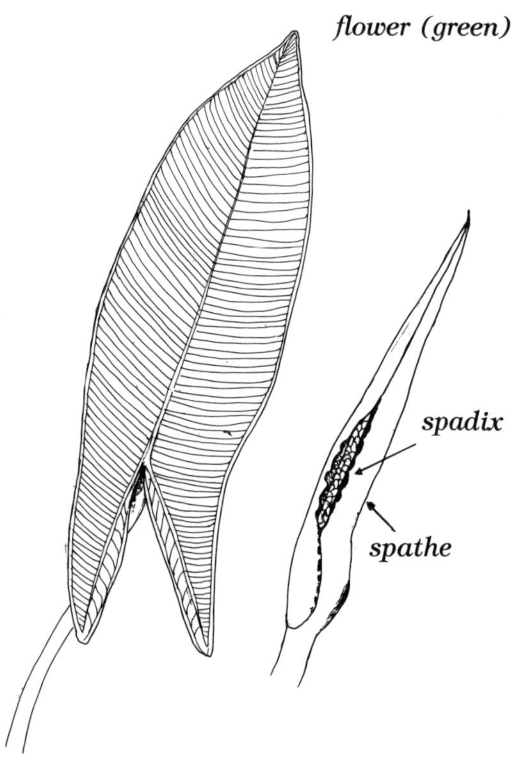

Arrow Arum

(Peltandra virginica)
Arum family

Size: 1-2′

Arrow arum is usually found on the margin of ponds and streams in shallow water. It prefers muddy soil.

This emergent has leaves that are usually arrow-shaped. Because of this, it is sometimes confused with arrowhead, another emergent often growing in the same habitat.

One of the big differences between these two plants is the flower. Arrow arum has a flower similar to that of skunk cabbage and Jack-in-the-pulpit. It consists of a green sheath-like spathe and a white finger-shaped spadix. This structure contains both male and female flowers . . . female at the bottom and male at the top. Arrowhead has flowers that are white and they rise above the plant in whorls.

The fruit of arrow arum are green berries with one to three seeds. They are a favorite food of wood ducks and are occasionally eaten by other birds, but the animal life found in the water around the plant, probably has more value as food for ducks than the berries. Indians used the root as a source of food, but it had to be cooked first to break down the crystals of calcium oxalate, making the plant poisonous if eaten raw.

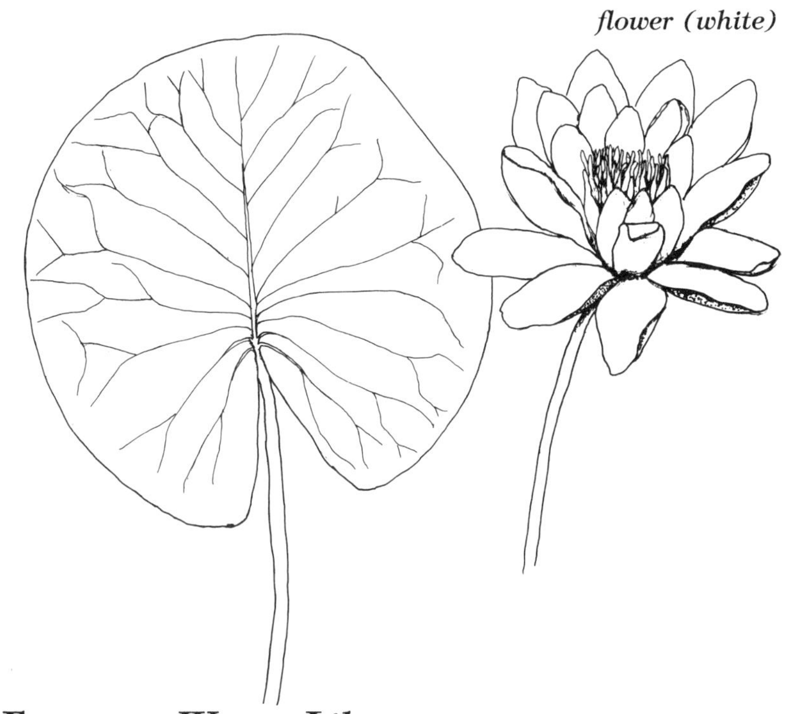

Fragrant Water Lily

(Nymphaea odorata)
Water lily family

Size: 4-12″ (leaves)

There are several species of water lily that are found growing in the Northeast, both yellow and white. None are more attractive than the fragrant white water lily *(Nymphaea odorata).*

It has round floating leaves that are purple underneath. The three-to-five inch flower is usually white, but may also be pink and is extremely fragrant. A very similar species is the tuberous white lily *(Nymphaea tuberosa).* This latter species has leaves that are green underneath, and the flower is not fragrant. The flower of both species open early in the morning, and usually start to close by mid-afternoon. On cloudy days, the flowers do not open. This probably helps to protect the pollen against rain.

The fragrant white water lily will grow in water as much as 10 feet deep, if the water is clear. It usually grows in ponds and slow streams where the water level doesn't change. Mats of these lilies sometimes crowd out other species more valuable to wildlife.

Water lilies are often used as an ornamental throughout the world, but they have limited value to wildlife. The young leaves may be eaten by fish, birds and muskrats and the seeds are eaten by ducks and other waterfowl to a limited degree.

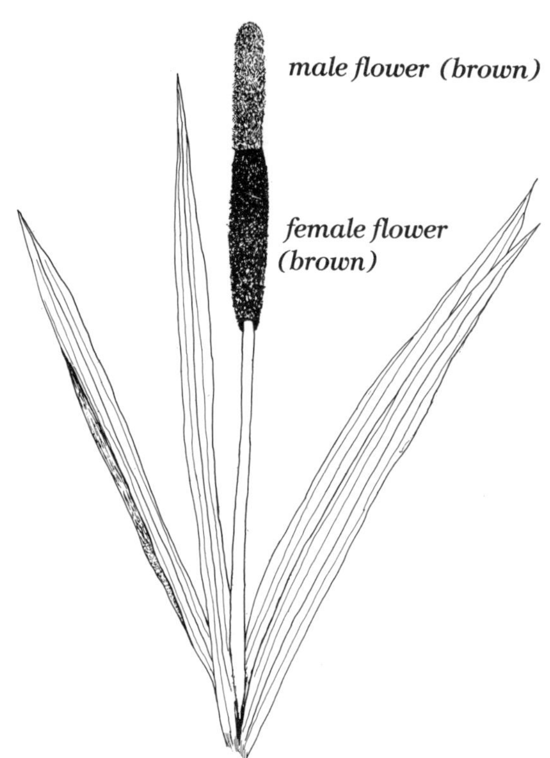

Common Cattail

(Typha latifolia)
Cattail family

Size: 3-9′

This well-known rooted emergent plant may grow along the pond's edge. It also grows in wet meadows and other wetland sites, often serving as an indicator of an area that has been disturbed by human activities.

The common cattail has rigid upright leaves and the most conspicuous feature is its flower with its well-known, cylindrical spike. This is made up of two portions; the upper part consists of the staminate flowers, and the lower section the pistillate flowers. The pistillate flowers persist and become the fruit containing thousands of tiny, hairy, wind-transported seeds.

Cattails provide excellent cover for nesting waterfowl and a number of songbirds, such as marsh wrens and red-winged blackbirds. It offers little in the way of food for most waterfowl, but the stems and rootstalks are eaten by muskrats who also use the stalks to build their houses.

Man uses the common cattail in many ways. The sprouts at the top of the rootstalk are sometimes eaten as greens or in salads. The leaves have been used for making rush-bottom furniture and the fluffy fruit has been used as insulation in blankets. At one time, paper was made from cattails. Today, man is again experimenting with this practice.

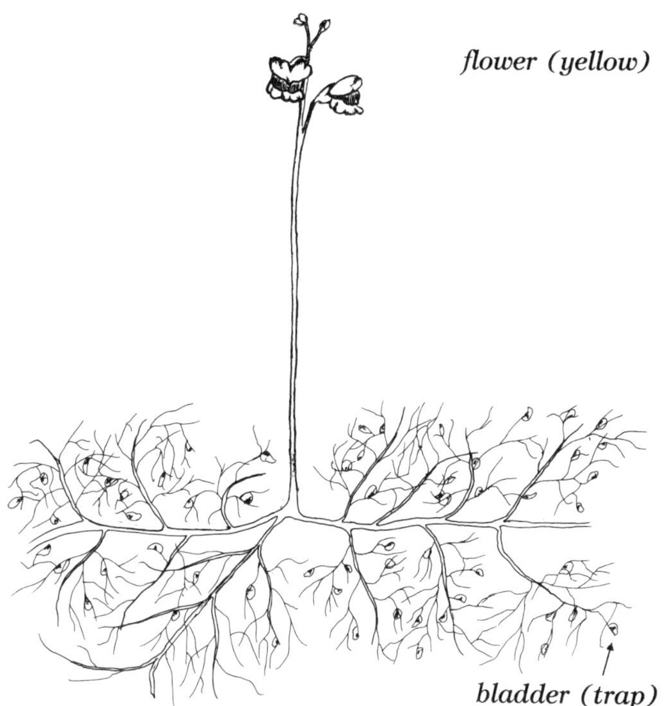

Common Bladderwort

(Utricularia vulgaris)
Bladderwort family

Size: $2^1/2$-8″ (flower stalks)

The common bladderwort is a carnivorous submerged plant unattached to the substrate. The genus name, *Utricularia,* comes from the Latin word utriculus meaning "small bottle." This refers to its many small bladder-like traps that are attached to the plant. It is often overlooked because all parts of the plant are submerged, except its yellow snapdragon-like flower. This extends above the surface of the water and blooms from June to August.

Common bladderwort is found growing in quiet waters that are either shallow or deep. It lacks roots, but has a branching stem from which arise many whorls of tiny green, sometimes divided branches bearing numerous tiny bladders. These are actually modified leaves. They help to anchor the plant and serve as traps to catch small aquatic animals that are digested to aid in nourishing the plant. Each bladder is equipped with tiny, branched hairs around the opening to the trap. This trap is sprung when small organisms disturb these hairs. The trap can be reset to capture more prey in about twenty minutes.

These bladders vary in size. Some of the larger ones are able to catch organisms such as fish fry, water fleas and mosquito larvae, and sometimes even small aquatic plants as duckweed *(Lemna)* and watermeal *(Wolffia).*

Pickerelweed

(Pontederia cordata)
Pickerelweed family

Size: 1-2′

Pickerelweed is often found along the edge of a pond growing in shallow water, but it will grow successfully in water up to three feet in depth.
The flower of pickerelweed is a colorful purplish-blue spike, similar to the purple loosestrife, having three types of flowers. Each type has six stamens and one pistil, but the length of these parts varies in each of the flower forms to ensure cross-pollination. The flowers that are most productive of seed appear to be those where cross-pollination involved stamens and pistils of equal length. It blooms from July to September.

The leaves of pickerelweed are usually somewhat heart-shaped but may vary in appearance. They are supported by long stalks.

There is often an abundance of small aquatic animals in and around this plant. A common game fish, the pickerel, is often found here, giving the plant its common name. Despite the fact that pickerelweed is a common aquatic plant, its value to wildlife appears to be limited. The seeds are eaten by black and wood ducks and muskrats, while deer may browse on the leaves. The seeds can be eaten like nuts. The Indians roasted them and added them to cereals and breads. They also ground them and used them as flour.

Chapter 10

Plants of a Freshwater Pond Community

Arrow Arum - *Peltandra virginica*
Blue Vervain - *Verbena hastata*
Canada Rush - *Juncus canadensis*
Cardinal Flower - *Lobelia cardinalis*
Common Arrowhead - *Sagittaria latifolia*
Common Bladderwort - *Utricularia vulgaris*
Common Cattail - *Typha latifolia*
Cow-Lily - *Nuphar advena*
Ditch Stonecrop - *Penthorum sedoides*
Duckweed - *Lemna* sp.
Floating Hearts - *Nymphoides cordata*
Fragrant Pond Lily - *Nymphaea odorata*
Least Spike Rush - *Eleocharis acicularis*
Pickerelweed - *Pontederia cordata*
Poison Sumac - *Toxicodendron vernix*
Pondweed - *Potamogeton* sp.
Purple Loosestrife - *Lythrum salicaria*
Red Maple - *Acer rubrum*
Sallow Sedge - *Carex lurida*
Seedbox - *Ludwigia alternifolia*
Skunk Cabbage - *Symplocarpus foetidus*
Small Water Plantain - *Alisma subcordatum*
Soft Rush - *Juncus effusus*
Spicebush - *Lindera benzoin*
Swamp Azalea - *Rhodendron viscosum*
Sweet Pepperbush - *Clethra alnifolia*
Tuberous Water Lily - *Nymphaea tuberosa*
Watermeal - *Wolffia* sp.
Water Plantain - *Alisma subcordatum*
Water Purslane - *Ludwigia palustris*
Water Shamrock - *Marsilea quadrifolia*
Water-shield - *Brasenia schreberi*

11

Flowers of the Seashore

THE WORLD OF the seashore is filled with beauty and discovery at every turn and combined with all this is the characteristic smell of the seashore and the persistent cry of the gulls. The bird life and the many invertebrates that adapt themselves so well to the intertidal zones are a separate study and all part of the interrelated whole, but the flowering plants of the seashore by themselves offer a rich field for exploration.

Each seashore will have its own mixture of flowering plants depending on the circumstances. Some seashore areas will have long stretches of sand or cobble beaches, others will have salt marshes. There may be a combination of all of these environments that are in close proximity to each other. Introduced trees and shrubs may add their special appeal along some shorelines. Many of our common roadside plants may also be present, which is not surprising because they are plants of waste or disturbed areas such as parking areas and dikes that are constructed to protect against coastal storms.

The place that we will be visiting has many of the features mentioned above. We approach the salt water by driving over a causeway connecting what was once an island to the mainland. There is salt water on either side of us.

One of the first plants to attract our attention growing along the causeway is the salt spray, or wrinkle-leaved rose. Most of the blossoms are deep pink, but some may be white. It is characterized by having heavily spined stems, compound leaves with wrinkled leaflets and a fruit (the hip); these are about an inch in diameter and high in vitamin C. This species has an interesting history having arrived on Cape Cod in 1849 as the result of a shipwreck. Crates of this plant were washed ashore and became established. It has since spread widely along the shore and inland. Some years ago, I discovered some wrinkle-leaved rose growing in a garden in northern New Hampshire.

Another interesting plant growing near the roadside is tamarisk, or salt cedar, as it is sometimes called, probably because it looks as if it would be an evergreen although it sheds its leaves. It has tiny pink flowers that bloom during the latter part of May or early June. It seems to do well at the seashore, but it also grows well inland. It has escaped from cultivation in the East and is found north to Massachusetts and Indiana.

The time to visit the shore is at low tide, when we can see the lush stands of cord grass, both tall and short cord grass. The species that we see here is mostly the tall cord grass growing closest to the water. Both species play a dominant role in salt marshes as they are among the few plants that help hold the marsh in place. Later, we will observe the plant zonation in a salt marsh and more will be said about the cord grasses.

As we drive farther, we come to buildings, branching roads, parking areas and a swimming beach and a picnic area. Along this route, we find such trees as the Russian olive, an ornamental with narrow, light gray-green leaves, the red cedar, black cherry and tree-of-heaven, all plants that move in as pioneer species in disturbed areas or spots that are no longer used.

Much of the vegetation around the seashore is like the plant life that we find farther inland. It is only when we leave our car and move closer to a salt marsh that we begin to notice a definite zonation. This is an indication of the effect that the tides and the salt concentration have on the distribution of plants near the shore.

Here, there is first an intermediate zone, a region between the land and the salt marsh. The plants that are found here are those that can adapt to the salt spray, but not to salt that would come in contact with their roots. Common species are switch grass, bayberry, poison ivy, growing as a shrub, and marsh elder, or high water shrub. Other border plants are seaside goldenrod, the lovely sea lavender and swamp rose mallow that brings spectacular beauty to the marsh in late summer when it is in full bloom.

We are now on the fringes of the upper marsh that is only submerged during very high tides. This is a fine opportunity to see the kinds of plants that make up the salt marsh itself. Because conditions become harsh as we get closer to the shoreline, the plants found here must be able to adapt well to a variety of environmental conditions. Chief among these adaptations is the ability to cope with varying exposure to water and salt. They face much the same problems as desert plants. There must be adaptations to help conserve on water and some way of keeping most of the salt out of their tissues or to get rid of excessive salt that they may take in.

An important plant found in the higher or upper marsh is the short cord grass. The genus name *Spartina* is a Greek word meaning "cord." This refers to a European species that at one time was used to produce cord. Short cord grass is the source of salt marsh hay used by early settlers along the coast. Mixed with the short cord grass is usually spike grass, both of which look much alike until they flower, and black grass. The latter plant is not a grass, but a rush. This mixture of plants varies from one salt marsh to another with the short cord grass usually being the dominant species.

Along the path leading to the lower marsh and in disturbed areas along the edge of the marsh, there are several species of pigweed, such as glasswort, sea-blite and orach. They seem to be very prevalent in this type of environment. Perhaps the most unusual looking of these plants is glasswort, once used as a substitute pickle by the early colonists, probably selected because of its resemblance to European samphire, used as a pickle in Europe.

Closely resembling glasswort is sea-blite, a plant that is somewhat succulent with narrow pointed leaves that are grayish green. Sea-blite can be eaten as a cooked green.

Orach has triangular leaves similar in color to sea-blite and is another wild edible. Here, it grows in open spaces near the edges of a salt panne, a small pool with a high concentration of salt.

Perhaps the most delicate, and beautiful plant found in this portion of the marsh is the salt marsh sand spurrey, a member of the pink family. The flowers are a few inches high and are pink or white. They open only during part of the day, so on some occasions they may be overlooked because of their diminutive nature.

A long dike separates the upper marsh from the lower portion of the marsh. On the side of the dike closest to the water, large boulders have been placed to stop the ocean water during extremely high tides. The side of the dike away from the water has been packed with soil. This allows plants like mugwort, lamb's-quarters, burdock and many other alien plants to take over. The top of the dike is flat, to allow for a foot path for hikers. This path offers a good opportunity to observe several species of grass such as foxtail, stink grass, goose grass, and brome grass. There are also plantings of beach plum and autumn olive to attract wildlife. Autumn olive is related to the Russian olive, but has leaves that are darker green and more egg-shaped. Both species have silvery scales on the undersurface of their leaves.

The lower marsh consists of predominantly tall cord grass, but in some disturbed areas, there are large patches of sea lavender, a member of the leadwort family. The tall cord grass grows from three to nine feet in height and at low tide, salt crystals can often be observed on its leaves. This is excess salt that has been excreted from glands on its leaves.

A view of clumps of sea lavender seen at a distance makes a lovely picture, and the individual flowers have their own special beauty when seen with the hand lens.

We leave the salt marsh and move to look briefly at another environment . . . the cobble beach. Plants grow higher on the beach where the soil is sandy. The rocky, cobble part of this beach lacks vegetation.

A few of the common flowers growing here are the jimsonweed with its purplish, funnel-shaped flower, the sea-burdock and the prickly saltwort, closely related to the Russian thistle. The leaves of the saltwort are spine-tipped, an adaptation for deflecting the sun's rays, hence cutting down on the evaporation from the leaves. Sea-burdock is well named because of its burr-like fruit. The rest of the vegetation consists of many pigweeds.

Further back, near the road, we find tamarisk and salt spray rose again and a new plant . . . sea rocket. This is a mustard easily recognized by its four petaled flowers and fleshy, rocket-like fruit.

Groundsel, beach pea, salt-marsh fleabane, beach grass and slender fragrant goldenrod are a few plants that are not found at this particular seashore, but may be common elsewhere.

We leave the beach to return to our cars. On the way, we can notice many plants typical to disturbed areas . . . sow-thistle, black knapweed, prickly wild lettuce, seabeach dock, black medick, burdock, prostrate knotweed, trailing wild bean, several species of plantain, milky purslane, rabbit-foot clover and pineapple weed, so-called for its odor.

As you will have seen, all seashore plants have their own distinctive features and adaptations, thus contributing to the excitement of seashore botanizing. Perhaps you will discover these and more for yourself when you next visit the shore.

Some Salt Marsh Plants

Saltwater Cordgrass

(Spartina alterniflora)
Grass Family

Saltmeadow Cordgrass

(Spartina patens)
Grass Family

Spikegrass

(Distichlis spicata)
Grass family

Blackgrass

(Juncus gerardi)
Rush family

Swamp Rose Mallow

(Hibiscus palustris)
Mallow family Size: 4-8′

Swamp rose mallow is a showy perennial with many stout stalks. The mallows belong to a family characterized by having many stamens that join to form a somewhat tubular enclosure around a five-parted style. The style extends beyond this sheath and has knob-like stigmas. Many pollen-bearing anthers appear at right angles to this tube. Familiar members of this family are hollyhock and Rose of Sharon.

Swamp rose mallow grows in salt marshes along the Atlantic coast from Massachusetts to Florida and in some fresh water marshes. This flower is probably our most spectacular marsh flower with its four to seven inch corolla that flares outward. It blooms from July to September and is usually pink, although sometimes it is white or purple. There is a closely related plant *(Hibiscus moscheutos)*, that is very similar with flowers that may be white or pink with red or purple centers. Both are called swamp rose mallow.

The leaves of *Hibiscus palustris* are egg-shaped, toothed, pointed and approximately four inches in length. The fruit is a capsule with five sections that separate at maturity. Each division contains several seeds.

Crushed leaves of swamp rose mallow were once mixed with honey and applied to the eyes to remove inflammation.

Sea Rocket

(Cakile edentula)
Mustard family

Size: 6-12″

Sea rocket is one of a small group of plants that is able to tolerate the salt spray that it encounters on the sand dunes. Its succulent nature is an indication of its ability to store water, a useful adaptation for combating the harsh conditions present along a sea coast.

This species is found growing on sand or gravelly soil on the Atlantic coast from New Jersey to Labrador. It is also found along the Pacific coast and along the Great Lakes.

Sea rocket is a rather sprawling, branched annual with fleshy stems and leaves that are both peppery to the taste. Its leaves are wedge-shaped at their base and may vary in length from two to five inches. The species name *edentula* means "toothless," referring to the leaves, although they may have wavy, toothed margins.

This member of the mustard family has very small pink to lavender, four-petaled flowers, 1/4 of an inch wide. It blossoms from July to September.

The seed pods resemble a tiny rocket ready to be launched. It consists of two sections with one seed in each part. Sometimes the seed in the lower section is missing. The whole seed pod is about an inch long.

The stem, leaves, flowers and seed pods are all edible. They are used both in salads and cooked as greens.

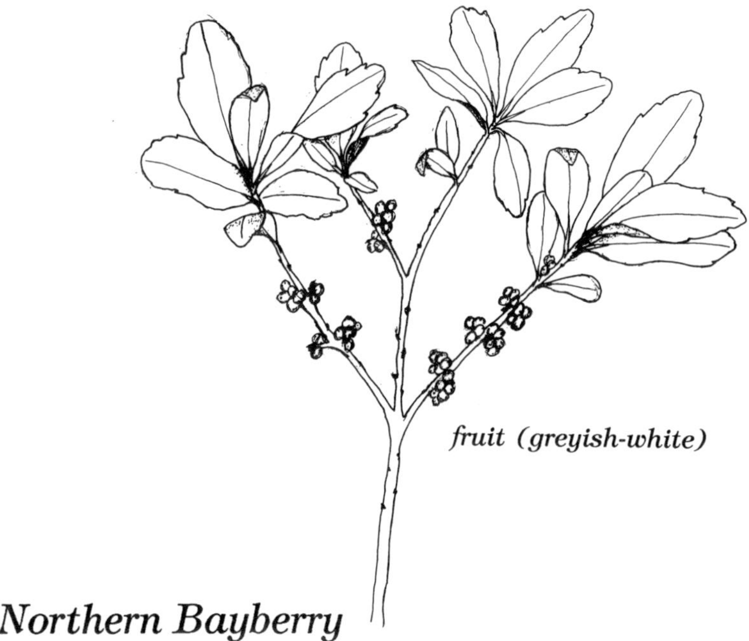

Northern Bayberry

(Myrica pensylvanica)
Wax-Myrtle family

Size: 3-6′

Northern bayberry seems to thrive in a salt water environment as it is able to take the saltspray from storms; however, it is unable to tolerate salt at its roots, so it grows on sand dunes some distance from the low tide region.

This species is found growing along the Atlantic coast from Nova Scotia to Florida. It also grows inland on sandy sites.

Northern bayberry usually grows as a shrub, but can grow to a height of 35 feet. Its flowers are in the form of catkins. These catkins are wind-pollinated and the sexes are usually present on separate plants. The flowers bloom from May to July while clusters of grayish-white waxy berries appear in June on some specimens. The fruit may sometimes be persistent on the plant for several years.

The leathery leaves of the northern bayberry are from two to five inches long and are covered with a scattering of golden resinous dots.

Bayberry is perhaps best known for its early use in making candles out of the wax derived from the berries. However, it has had many other uses; i.e. as a dye, a seasoning and medicinally.

Many birds feed on the berries, especially the tree swallow during migration and the yellow-rumped warbler during the winter months.

Beach Plum

(Prunus maritima)
Rose family

Size: to 8′

Beach plum is usually a rather low-growing, sprawling shrub, but its growth habit is quite variable. Sometimes it may lie prostrate close to the ground, while occasionally it grows as an upright tree. The latter is true particularly when beach plum grows inland, away from the shore.

This member of the rose family is usually found growing in sand dunes along the Atlantic coast from New Brunswick to Virginia. It generally is found along the seashore, but appears to do well in the New Jersey pine-barrens and in the Indiana sand-dunes.

The beach plum has very attractive white flowers that appear in dense umbels from April to June. The stem is densely branched and often crooked.

The beach plum's leaves are oval and pointed. They vary in size from 1 to 2$^{1}/_{2}$ inches and have many small teeth. The fruit ripens from September to October. It is fleshy with a cherry-like stone, and is usually about $^{3}/_{4}$ of an inch in diameter. The color of the fruit is usually purple, but some forms may be either red or yellow.

Beach plum is considered a choice species for use in making pies, jellies and jams. It is also a valuable plant for controlling beach erosion.

Common Glasswort

(Salicornia europaea)
Goosefoot family

Size: 2-16″

The generic name *Salicornia* is derived from two Latin words meaning "salt horn." This is quite descriptive of its many cactus-like branches. These branches are jointed and are much longer than they are wide. The scale-like leaves are greatly reduced in size, an adaptation for conserving water. Its flowers are very small and inconspicuous. They form from July to November in the spaces between the upper joints of the fleshy stem. The common glasswort is an annual that appears singly and is soft throughout its structure.

Another species of *Salicornia* that may grow in the same area is woody glasswort *(Salicornia virginica)*. It is a perennial with creeping stems that have a woody center, and it commonly forms mats.

The common glasswort does best when it grows in open or disturbed areas in a marsh, where it is only occasionally flooded by the tide. Its seeds germinate easily in this environment where the salt content is high.

The early colonists used this species for making pickles and added the branch tips to salads to give them a salty flavor. The chemical make-up of *Salicornia* is also useful in glass making. Ducks and geese feed on the seeds and tender branches, particularly the pintail ducks.

Chapter 11

Flowers of the Seashore

Autumn Olive - *Elaeagnus umbellata*
Beach Grass - *Ammophila breviligulata*
Beach Pea - *Lathyrus japonicus*
Beach Plum - *Prunus maritima*
Black Cherry - *Prunus serotina*
Black Grass - *Juncus gerardi*
Black Knapweed - *Centaurea nigra*
Black Medick - *Medicago lupulina*
Bouncing Bet - *Saponaria officinalis*
Brome Grass - *Bromus tectorum*
Burdock - *Arctium* sp.
Common Glasswort - *Salicornia europaea*
Foxtail - *Setaria* sp.
Goose Grass - *Eleusine indica*
Groundsel - *Baccharis halimifolia*
Jimsonweed - *Datura stramonium*
Lamb's-quarters - *Chenopodium album*
Marsh Elder - *Iva frutescens*
Milky Purslane - *Euphorbia supina*
Mugwort - *Artemisia vulgaris*
Northern Bayberry - *Myrica pensylvanica*
Orach - *Atriplex patula*
Pineapple Weed - *Matricaria matricarioides*
Plantain sp. - *Plantago* sp.
Poison Ivy - *Toxicodendron radicans*
Prickly Saltwort - *Salsola kali*

Prickly Wild Lettuce - *Lactuca canadensis*
Prostrate Knotweed - *Polygonum aviculare*
Rabbit-foot Clover - *Trifolium arvense*
Red Cedar - *Juniperus virginiana*
Russian Olive - *Elaeagnus angustifolia*
Salt-marsh Fleabane - *Pluchea purpurascens*
Sand Spurrey - *Spergularia marina*
Seabeach Dock - *Rumex pallidus*
Sea-blite - *Suaeda maritima*
Sea-burdock - *Xanthium echinatum*
Seaside Goldenrod - *Solidago sempervirens*
Sea Lavender - *Limonium nashii*
Sea Rocket - *Cakile edentula*
Short Cordgrass - *Spartina patens*
Slender Fragrant Goldenrod - *Solidago tenuifolia*
Sow-thistle - *Sonchus oleraceus*
Spike Grass - *Distichlis spicata*
Stink Grass - *Eragrostis* sp.
Swamp Rose Mallow - *Hibiscus palustris*
Switch Grass - *Panicum virgatum*
Tall Cordgrass - *Spartina alterniflora*
Tamarisk - *Tamarix gallica*
Trailing Wild Bean - *Strophostyles helvola*
Tree-of-heaven - *Ailanthus altissima*
Woody Glasswort - *Salicornia virginica*
Wrinkle-leaved Rose - *Rosa rugosa*

12

The Goldenrods

GOLDENRODS, ALONG WITH other composites, such as asters, Joe-Pye-weed and boneset, all herald the approach of autumn. They are one of the most widespread wildflowers in America, particularly in the eastern section of the United States. These perennials are chiefly a North American genus. England, for example, has only one species – *Solidago virgaurea,* commonly called woundwort. Many goldenrods are common along roadsides and dominate some old fields that have not recently been cultivated, while other species are found in woodland, wetland and seashore areas.

The numerous small flower heads of the goldenrod are arranged in a variety of ways, but both ray and disk flowers are present and are usually yellow. There may be variations in some species where there are white ray flowers and yellow disk flowers. The common silver-rod is the only goldenrod species that has both ray and disk florets that are white.

The ray florets are not evenly arranged around the central disk as is true of asters, and the number is far fewer, usually varying from one to twenty.

Goldenrods, as a group, are very easy to recognize and some species can be readily learned without confusion. However, there are approximately 125 species growing in the United States and there are also many varieties. The tendency of many of the species to hybridize is one of the factors that makes mastery of many of the goldenrods so difficult.

One approach that simplifies the study of goldenrods is found in *A Field Guide to Wildflowers* by Peterson and McKenny. This book is in the well-known Peterson Field Guide series. This method makes use of the silhouette of the plant to place the goldenrod in one of several groups. You are to decide if the goldenrod you are looking at has plume-like, elm-branched, club-like, wand-like or flat-top flower clusters. You are next asked to observe if the leaves have parallel or feather-veining.

With a little practice it will probably be possible to identify at least a dozen species without too much difficulty. The problems arise when several species of goldenrod look very similar, such as the Canada goldenrod, the late goldenrod and the tall goldenrod which are all common to roadsides and fields. All three may be tall (up to five feet or over) with plume-like flowers. Their leaves have toothed margins, all similar, having three parallel veins. It is only by noting small differences in these plants that an identification is possible.

Even experts cannot always agree on how these three plants should be classified. However, there are several goldenrods that are relatively easy to identify.

Our trip to look at goldenrods is to a spot along a roadside, including fields and woods. We disperse to varying habitats to see how many different kinds we can find. This method could be repeated at other sites if you are interested in learning more than a few of the common species. Any specimens that cannot be readily identified can be collected and placed between newspaper.

If there is an interest in mounting any of these specimens for an herbarium, the stem may be folded into a V or N where the plant is extremely long (some goldenrods grow to be six to eight feet tall). The mid-section can be removed leaving the upper and basal parts.

One of the more easily identified goldenrods is the rough-stemmed goldenrod, which can be identified by its wrinkled, hairy stem and its wrinkled, hairy, deeply-toothed, feather-veined leaves. Its flower cluster is plume-like. Look for it in thickets, roadsides and open fields.

Zig-zag goldenrod has a zig-zag stem with very broad and well-pointed leaves at both ends. The flowers are in small clusters in the upper leaf axils and at the summit. This is a plant of woodlands and thickets.

Blue-stemmed goldenrod has flower tufts in the axil of its smooth bluish or purplish stem with a white bloom. Like the zig-zag goldenrod, it is usually found growing in woods and thickets.

Rough-leaved goldenrod has elm-like branching supporting its flower clusters and very harsh, rough leaves. The lower leaves are very large and tapering to the base. Its stem is sharply four-angled. The habitat of the rough-leaved goldenrod is usually wetland areas such as swamps, bogs and wet meadows. Additional species of goldenrod will be treated in greater depth at the end of the chapter.

Goldenrods, like other composites, have a special adaptation to avoid self-pollination. This is accomplished by the anther and stigma maturing at different times. In goldenrods, the pollen ripens before the stigma matures.

A great many insects visit goldenrods. They include honeybees, bumblebees, syrphid flies, carpenter bees and many more. Each goldenrod flower secretes only a small amount of nectar, but the presence of over a 1000 flowers in each goldenrod plant compensates for this.

Bumblebees are one of the regular visitors to goldenrods. An interesting behavior can sometimes be seen when large and small bumblebees (two different species) are foraging. The large bumblebee feeds on the flowers closest to the stem, while the small bumblebee feeds on the outermost flowers. These flowers will not support the heavy weight of the large bumblebees. In this way they avoid competing with one another.

Goldenrods have had a very long and interesting history and have had many uses. The name Solidago comes from the Latin word *solido* meaning "to make whole." This undoubtedly was based on the medicinal qualities of this plant. Its use as a wound-herb goes back at least to the time of the Crusades. At that time, it was known as the Saracen's wound-herb. Eventually, the Crusaders brought the plant to Europe. Later, in England, and also in China, it was to be very highly regarded as a plant that could control bleeding from wounds. The common name "woundwort," given also to the one species of goldenrod native to England, alludes to its healing powers. It was useful for treating all kinds of sores and ulcers, particularly those of the throat and mouth. One preparation was even said to fasten teeth that were loose in the gums.

During the Elizabethan period, the English paid large sums for great amounts of goldenrod that came from America. In the early part of the 1600's woundwort was found growing in England. This sent the price of goldenrod plummeting.

In America, the Indians and colonists depended heavily on goldenrod. They used it internally to reduce fever, to treat bladder and kidney ailments and as a general tonic. Externally, it was used for headaches and as a poultice for boils, snake bites, ulcers and other bleeding. As a medicine, all parts of the plant were used.

Following the Boston Tea Party, the colonists used a substitute called Liberty tea. The main component was from sweet goldenrod. Another tea made from sweet goldenrod is known as Blue Mountain tea. At present, it is sold by many health food stores.

The flowers, either fresh or dried, can be used by the cupful in such foods as bread, fritters and cake batters. This makes a tasty addition. The dried blossoms were also used in tobacco mixtures. Goldenrod is an excellent, but little used source of a yellow dye.

At one time, powdered goldenrod flowers were used to induce people to sneeze. It was carefully administered and was considered a good joke. This may be the origin of the belief of many that goldenrod causes hay fever. The real source of this irritation is ragweed. They both blossom about the same time, but ragweed is wind-pollinated and goldenrod is largely pollinated by butterflies and other insects.

Another interesting belief was associated with the swellings (galls) found on goldenrods. There are several types of swellings made by gall midges, flies or moths depending on the species. These galls were referred to as "rheumaty buds" and were often carried around in one's pocket. It was believed that there would be relief from the pain of rheumatism as long as the insect developing inside the swelling was present.

This reminds me of my childhood experiences with galls. I, along with my friends, would collect the dried stalks of goldenrods with the round swellings of the goldenrod ball gall present, located usually part way down on the stalk. We would break off the top part of the stalk above the gall and use the gall to hit each other over the head. We called them "noggin knockers."

Perhaps you can find a patch of goldenrod near your home and watch it over a period of time. See what insects visit it, and notice how the flowers change and the fruit forms during the autumn months.

Gray Goldenrod

(Solidago nemoralis)
Aster family

Size: 1 - 3′

Gray goldenrod, or dwarf goldenrod as it is sometimes called, is usually rather easy to identify. This is because of its diminutive stature, the presence of tiny leaflets in the axils of the leaves and a grayish stem. It is a slender species, less than a foot in height, but occasionally grows to be over 4 feet tall.

Gray goldenrod is commonly found growing where the soil is dry and shady. This includes fields, old pastures and open woods. The species name *nemoralis* means "of woodland."

The stem of the gray goldenrod develops from a branched rootstalk, and is covered with dense, fine gray hairs. These are best seen with a hand lens. At the top of the stem, there are short, lateral branches ending in a narrow, but dense one-sided, plume-like cluster of flower heads. Each flower head is very small, being only $^1/_{12}$ to $^1/_8$ of an inch long. There are five to nine bright yellow ray flowers and three to six disk flowers. The flowers appear from August to October.

Its broad, basal leaves are stalked and toothed. The upper leaves are smaller, untoothed and without stalks. All of the leaves are rough due to the presence of hairs.

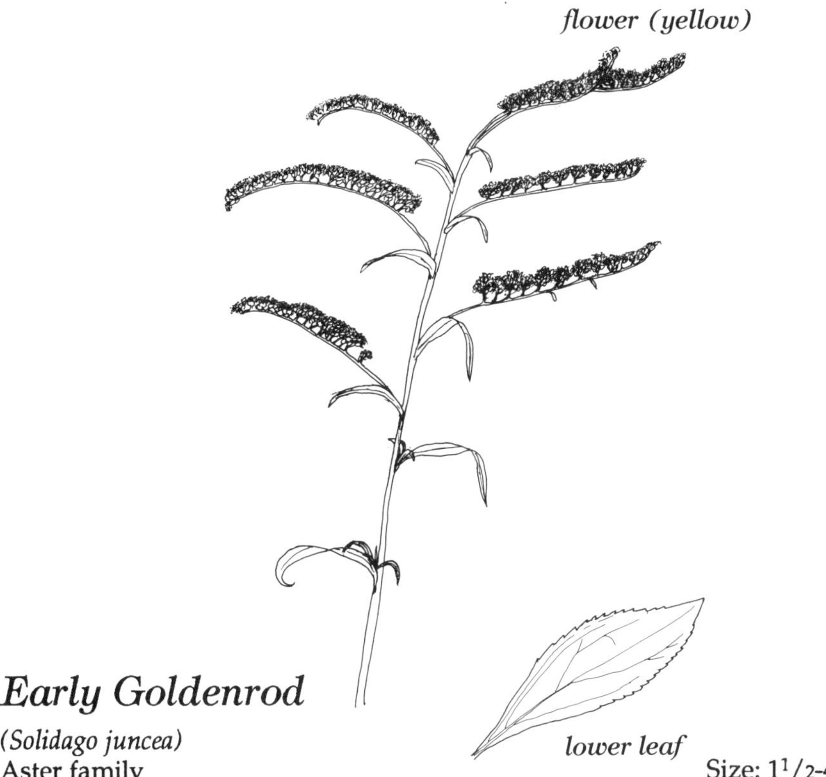

Early Goldenrod

(Solidago juncea)
Aster family

Size: $1^1/_2$-4′

This goldenrod is characterized by having large, broad, long-stalked basal leaves and the presence of many tiny leaflets in the axils of the upper leaves. The species name *juncea* means "with stiffness," similar to a rush. This doesn't seem to be too descriptive.

Early goldenrod grows in open, dry places such as fields, borders and open woods. It is a medium-sized goldenrod with stems that usually appear singly or a few in a group. These develop from a thick, branched rootstalk or sometimes from slender runners. The stem may be light green or brown and is hairless.

The leaves of the early goldenrod are smooth and sharply toothed, tapering to a hairy stalk. Its leafstalk encircles the stem by more than half the circumference of the stem. The upper leaves decrease in size upward with the upper leaves being toothless and quite narrow.

Early goldenrod has flower clusters that are usually plume-like but may be elm-like. The flower heads grow along one side of the several spreading branches which droop at the tips. This is the first goldenrod to bloom, flowering from mid-July to mid-August. There are 8 to 12 ray flowers.

The root of the early goldenrod was used by the American Indian to make a solution taken internally for the treatment of convulsions.

Canada Goldenrod

(Solidago canadensis)
Aster family

Size: 1-5′

Canada goldenrod is a common species but it closely resembles two other common goldenrods . . . late goldenrod *(Solidago gigantea)* and tall goldenrod *(Solidago altissima).* All three species have flower heads in clusters that are plume-like. They all have toothed, lance-shaped leaves with three prominent veins.

The characteristics of the Canada goldenrod that are different from the other two species are: the presence of small dense hairs on the stem just below the flower clusters, with the stem becoming hairless toward the base and without a pale purplish, waxy coating; the margins of the leaves are more deeply cut (often over one millimeter deep); and the flower heads are smaller (about 1/10 of an inch long). Late goldenrod has a smooth green or purplish stem, often with a whitish bloom, while the entire stem of tall goldenrod is covered with a grayish down.

Canada goldenrod grows along roadsides, in fields and in thin woods. The numerous yellow flowering heads blossom from July to September. There are 9 to 15 small ray flowers. Its leaves are very dense on the stem and are usually smooth. The upper leaves are stalkless and the lower ones stalked. They vary in length up to 5 inches, with a width of 1/2 inch.

The Indians used the flower tops to make a tea for relieving fever, and this was probably one of several species that they used for obtaining a yellow dye.

Flat-Topped Goldenrod

(Solidago graminifolia)
Aster family Size: 1-4′

The species name *graminifolia*, meaning "grass-leaved," aptly describes this common goldenrod. It is bushy in appearance, and is sometimes called lance-leafed goldenrod.

Flat-topped goldenrod is another goldenrod that has leaves that are fragrant when crushed. A similar but smaller species is the slender fragrant goldenrod *(Solidago tenuifolia)*; these also have grass-like leaves that are even narrower. Its habitat is usually along the edges of salt marshes.

The stem of flat-topped goldenrod is smooth or there may be fine downy hairs. The branches support a flat-topped group of small flower heads. Each head is about 1/5 of an inch long. There are from 10 to 35 ray flowers. The flowering period is from July to October.

This species lacks long-stalked basal leaves. The leaves are narrow, stalkless and sharply pointed with the lower leaves generally broader then the upper ones. Its largest leaves are about 1/4 of an inch wide. They have from three to five veins with occasionally as many as seven. The surface of the leaves has glandular dots. This is seen only when the leaf is magnified. A tea can be made from the leaves.

Sweet Goldenrod

(Solidago odora)
Aster family

Size: 1 1/2-3′

Sweet goldenrod is one of the few species of goldenrods that usually emits an odor when the leaves are crushed. This odor smells like anise, or licorice.

It has stems appearing singly or in small clusters. They are slender and generally smooth, but may be slightly hairy.

Sweet goldenrod has leaves that are numerous, thin, lance-like and quite narrow. They are usually stalkless, untoothed and vary in length from two to four inches, the upper leaves being very small. Each leaf has a prominent mid-rib beneath and its surface is covered with transparent dots.

It is generally found growing in dry, sandy soils in fields, open woods and clearings.

The plume-like flower heads droop slightly and are arranged along one side of the branches. They bloom from July to September. Each flower head is about 1/6″ long. The ray flowers vary in number from three to five.

Sweet goldenrod has been used in many ways over the years. The Indians made a solution from the flowers and used it for treating bee stings. Liberty tea, as previously mentioned, can be made from the dried leaves and flowers. This was once used to improve the unpleasant taste of certain medicines.

Chapter 12

The Goldenrods

Blue-stemmed Goldenrod - *Solidago caesia*
Canada Goldenrod - *Solidago canadensis*
Early Goldenrod - *Solidago juncea*
Flat-topped Goldenrod - *Solidago graminifolia*
Fragrant Goldenrod - *Solidago tenuifolia*
Gray Goldenrod - *Solidago nemoralis*
Late Goldenrod - *Solidago gigantea*
Rough-leaved Goldenrod - *Solidago patula*
Rough-stemmed Goldenrod - *Solidago rugosa*
Silver-rod - *Solidago bicolor*
Sweet Goldenrod - *Solidago odora*
Tall Goldenrod - *Solidago altissima*
Zig-Zag Goldenrod - *Solidago flexicaulis*

13

Wild Asters

EACH YEAR AS goldenrods and Joe-Pye-weeds begin to appear along roads and in fields and meadows, we become aware that summer is slowly slipping away. The time is fast approaching for a colorful, but somewhat more complex group of plants to appear. September and October are the months for the composites, considered the most advanced among our many flower families.

The purpose of this walk is not to make you an expert on composites, but to introduce you to this plant family and to show you how you can make a start, in tackling what many consider a very difficult group to study. I have always pledged to leave them alone, but have found the temptation to know more about them irresistible.

To start with, a few types stand out, and once you get the basic flower structure down and learn a few of the major differences among the various kinds of composites, particularly asters, you have taken a big step forward.

The composites are a very large and flourishing group of plants. This is probably because they are hardy and have many ways of adapting for survival, including special devices for seed dispersal. They have very compact flower heads with many florets in each head. This is what gives the family its name.

The flower head in composites is often mistaken for one single flower. The use of a hand lens will reveal that in all composites there are only two kinds of flowers, or florets, very closely packed . . . ray florets and disk florets. The ray florets are the colorful petals that are mainly for attracting the attention of insects. In some composites the ray florets are fertile, but they usually will have only the female parts. The disk florets in the center of the head have both male and female florets that are surrounded by a small, tubular, five-parted corolla that is jointed. This usually has five teeth. Disk florets may, like the ray florets, attract the attention of insects, but their primary purpose is to produce seed-like fruit . . . the achene.

The presence of the two kinds of florets in composites will vary. In sunflowers and asters, both types of florets are present. Chicory has only ray florets, while in burdock and thistle there are only disk florets.

Composites are also easily pollinated. An insect can readily make a trip to one flower head then travel to many individual flowers with the pollen it has collected. Asters cannot carry out self-pollination because the pollen will not germinate on its stigma. However, it will germinate on another aster.

On this walk we will look especially for asters. It is not difficult to learn many of the common types of asters by their size, the way the leaves clasp the stem, whether the stem is hairy or smooth and how branched they are. Some species are branched much more than others.

You may also need to look at small details, such as the appearance of the bracts, or modified leaves, positioned below the flowers; for this, a hand lens would be useful. Color in the same species may sometimes by quite variable, from white to blue to rose or purple. However, asters never have yellow ray flowers.

Habitat is also helpful in identification. Many species of aster are commonly found along roadsides, in clearings, in the woods and along the edges of forests.

Asters, like goldenrods, begin to bloom at the tip of the branches. The heads nearest the central stem bloom last. The ray flowers are often just female (pistillate). They develop a type of fruit known as an achene, these are small, dry, hard and one-seeded. The achene has a ring of pappus, or modified hairs, and this helps the wind to disperse it. The disk flowers are typically male and female. These are generally yellow, but in some species they are reddish or purplish, particularly as they age.

The word *aster* comes from the Latin and Greek word for "star." The remains of the flower can retain this star-like characteristic even in the winter.

There is a Greek legend that the aster was created from stardust. Wreathes of asters were considered sacred to all the gods and were put in temples on festive occasions.

We will begin our walk along a well-traveled country road that has various habitats along its length. These include the open environment along the roadside, the woodland edge and an open field. Within these varying environments, there will be an opportunity to observe asters of many colors, sizes and descriptions.

We scatter to collect from these varying habitats some of the brightly colored blooms around us; then gather to share our findings and examine our specimens more closely. Asters wilt quickly so you may wish to place them between newspapers if you want to preserve them for later examination.

The stiff aster can readily be identified by its many rough, dark green needle-like leaves that are crowded together on the stem; the smooth aster by its clasping, smooth, rubber-like leaves; the New England aster by its clasping leaves, large, deep purple flowers, hairy stem and glandular hairs on the flower stalk and flower bracts; the whorled wood aster by its stem that is often zig-zag with upper leaves so close together that they appear whorled, but are not.

A number of asters will have heart-shaped leaves. The large-leaved aster has basal, heart-shaped leaves that are rough and from four to six inches wide, bigger than any other heart-shaped species. The stem is usually angular and crooked and the floral branches have sticky glandular hairs. The white wood aster has heart-shaped leaves that are mostly stalked, and it has a smooth, slender stem.

One of the common asters you may find is the red-stemmed aster. A typical specimen will have purple flowers, a reddish, hairy stem with toothed leaves; however, some of these characteristics may be absent. If you think you have found this species, collect several specimens to see if they are typical or if they vary in some way.

It is relatively easy to identify the species I have mentioned. There are other asters that we may find, however, that will be quite confusing to identify. This is especially true of some of the small, narrow-leaved white asters, sometimes called frost asters. More work needs to be done on this particular group to make identification certain because even the experts cannot agree on what name to give to some species.

An example would be trying to determine the difference between *Aster pilosis* and *Aster ericoides*. Small white asters look so much alike that a very close look at fine details is necessary to distinguish one small white aster species from another, and then not always with certainty. This may mean observing the length of the leaves, whether they are toothed or untoothed, the number of ray flowers per head, the number of disk flowers per head and the appearance of the floral bracts.

In collecting and observing asters, it is important to note the absence or presence of basal leaves and if they are present, they should be included when the plant is collected. They are often an aid to proper identification.

Asters have been found to be useful in many ways. The leaves of some asters can be used as an herb, like spinach, if prepared when they are young and fresh. While a few species are poisonous, some provide good grazing for larger wild game animals and domestic animals such as horses and sheep. They also provide cover for many small wild game animals. All provide stability in holding the soil and all add beauty to their surroundings.

Stiff Aster

(Aster linariifolius)
Aster family

Size: 6-20″

This strikingly beautiful aster is also called savory-leaf aster. The species name *linariifolius* means "with linar-like leaves." This refers to the resemblance of its leaves to those of *Linaria* or toadflax a common roadside plant, also known as butter-and-eggs.

Stiff aster is a plant of dry sandy soil. It is occasionally found growing along roadsides. Because of its growth requirements, it is sometimes introduced into rock gardens that are sunny and dry.

The stiff, rough and somewhat hairy stem arises from a short rhizome, or underground stem. There are many crowded, toothless, narrow leaves on the stem. They are stalkless and stiff with rough margins, giving the feeling of sandpaper to the touch. The larger leaves are from 3/4 to 1 1/2 inches long and about 1/6 of an inch wide, with the upper leaves becoming modified into bracts on the stem of the flower stalk.

Stiff aster has a growth habit of appearing in little clumps or tussocks. The stalks are terminated by one or several heads of blue ray flowers with yellow disk flowers. They are up to one inch wide. The number of ray flowers varies from 10 to 20.

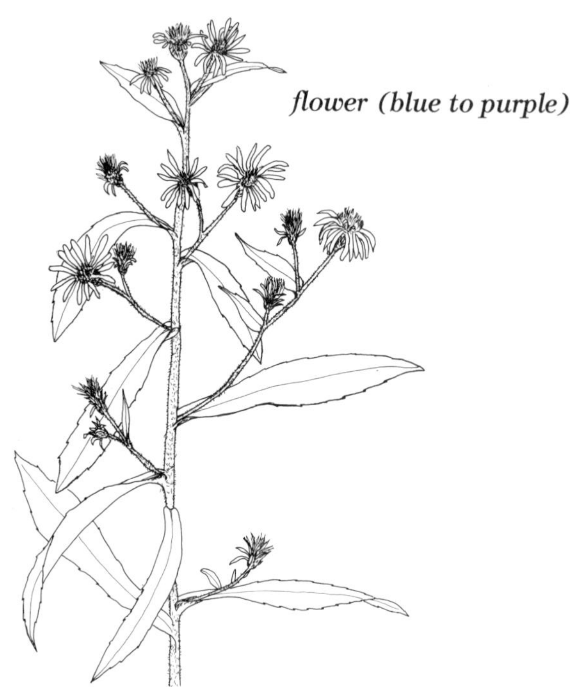

The Red-Stalked, or Purple-Stemmed Aster

(Aster puniceus)
Aster family

Size: 2-7′

This is a very beautiful but highly variable aster. It is usually found growing on rather wet sites, such as swamps, wet meadows and in moist spots along roadsides. Rickett, in his ***Wild Flowers of the United States*** *(Northeastern)* , reports walking through a wet area where this species was growing and observing that hardly two of these plants looked alike.

The secret to learning how to recognize this plant is to become familiar with its typical characteristics and then to observe other specimens that vary from the typical type.

The species name *puniceus* means "red or purple," describing the usual color of the stem, although it is sometimes green. In general, the stem is stout with bristly hairs, but at times it can be smooth.

The leaves are usually toothed and hairy but are variable in appearance, often being lance-shaped, and they clasp the stem at least a little. The size of the leaves range from about three to six inches.

The red stalked aster blossoms from August to October. Its flowerhead is nearly an inch in width. The number of ray flowers varies from 20 to 60 and the flower color may be blue to violet to rose, or even white.

Indians considered this an excellent medicine for treating nerve ailments.

The Smooth Aster

(Aster laevis)
Aster family

Size: 1-3′

The smooth aster is one of our loveliest blue asters. Its attractiveness has made it the source of many cultivars. This means that through plant breeding new varieties have been developed for use in gardens. Some varieties are even grown in the tropics.

The smooth aster has a stout stem with whitish powdery blooms on the branches. The stem arises from a short thick rhizome that may have additional creeping red rhizomes. Its leaves are succulent, very smooth and bluish-green in color, and feel almost like rubber to the touch. They are from one to four inches long, mostly untoothed, and often lance-shaped, but are quite variable. Leaves on the upper portion of the stem are stalkless and clasp the stem, while the lower leaves, including the basal ones, have stalks that are winged.

The violet-to-blue flower blossoms from August to October. It is nearly an inch wide. Each flower head has 15 to 20 ray flowers.

Indians once used the smooth aster to supply smoke for their sweat baths. It also provided smoke to revive people who had become unconscious, a paper cone being placed over the nose to pass the smoke through.

New England Aster

(Aster novae-angliae)
Aster family

Size: 3-7′

The New England aster is usually considered the most beautiful of our wild asters. This, and the New York aster, are ancestors of the Michaelmas daisy, popular in England, but not as well known in the United States. This species of aster has also been used for developing many of the "hardy asters" found in our gardens.

The New England aster is found in moist areas along roadsides, in fields, meadows and swamps. It has a thick stem covered with bristly hairs and several flower heads in clusters ending the branches. The leaves are crowded, lance-like and tightly clasp the stem and are longer than they are wide. They are untoothed and may be hairy and rough. Their length varies from $1^1/_2$ to 5 inches.

The flowers appear from August to October and are usually deep purple, but may vary from pink to lavender, and even to white. The flower heads are one to two inches across while the ray flowers are from 45-100 in number. Each ray is from $^1/_2$ to $^3/_4$ inch long. The stalks of the flowers are covered with an abundance of sticky, stalked glands. These are also present on the bracts of the flower heads.

Flat-Topped White Aster

(Aster umbellatus)
Aster family Size: 2-7′

The flat-topped white aster is a large bushy plant that prefers sites in sun or partial shade, but is found in a variety of habitats such as meadows, roadsides, edges of woods and swamps, and moist thickets. This species is a very good plant for the moist wildflower garden.

The stem of this species is usually smooth with many flower heads appearing on erect branches. Its leaves are lance-like to elliptical. They are toothless and up to six inches in length. The upper surface is dark green with the under surface being lighter. Margins of the leaf are rough.

The flat-topped white aster is one of the earliest asters to flower, blossoming from August to September. It is an easy plant to recognize because of its very wide, flat top cluster of flower heads. This top may reach a width of nearly a foot. Each head is from $^1/_2$ to $^3/_4$ of an inch wide and has 7 to 15 ray flowers.

The Indians used the flower of this plant to produce a smoke to combat evil spirits that might be interfering with the recovery of their sick. The dried leaves were used to make a tea.

Chapter 13

Wild Asters

Flat-topped White Aster - *Aster umbellatus*
Large-leaved Aster - *Aster macrophyllus*
New England Aster - *Aster novae-angliae*
Purple-stemmed Aster - *Aster puniceus*
Smooth Aster - *Aster laevis*
Stiff Aster - *Aster linariifolius*
White Wood Aster - *Aster divaricatus*
Whorled Aster - *Aster acuminatus*

14

A Touch of Autumn

ONE OF NATURE'S most spectacular displays is the autumn woodland in New England. It is as if an artist had dipped into his paint palette and gone slightly berserk. Gradually, but at the appropriate time, the forest becomes ablaze with a kaleidoscope of colors for all to enjoy. The deciduous forest, with its many species of trees and shrubs displaying somewhat different colors, seems to fit like pieces of a jigsaw puzzle into a picture of remarkable beauty.

We may get somewhat the same impression if we focus in on an isolated tree. For example, the sugar maple may have several colors present on the same tree. Notice the many colors present in the common trees in your neighborhood, and even in your own yard. These will usually range from the yellows of the birches and aspens to the reds and purples of many of the oaks. All of this color is due to the chemical break up of various pigments in the leaf as the tree begins to prepare for winter. The yellow pigments are masked by the green pigment, chlorophyll, that the plant uses for making food. Other pigments, especially the reds, form in the fall.

At the same time that this is happening, the stalks of leaves are gradually releasing their hold on the tree branch. One by one, they begin drifting to the street or forest floor.

Millions of Americans watch eagerly for the foliage to reach its peak. This usually varies by a few weeks in different parts of the Northeast. After the peak of color is past, we can see for the first time the bare "skeleton" of the woods before winter sets in. Now is the time to get to know and recognize individual trees and shrubs.

As the leaves fall from the trees, winter buds, tree bark, as well as the characteristic silhouettes of the various trees all become more obvious to our eyes. These characteristics can also be observed in winter, but often the snow obscures some of the details that stand out so vividly in autumn. Now, berries, acorns and fruit of every kind strike the eye . . . even the flowers of a late-blooming shrub, the witch hazel.

A walk through the woods is a special experience at any time of the year. In the late autumn it is unique. The leaves of every hue have by now drifted earthward to become a carpet, crackling underfoot. The branches of the trees are revealed to our eyes, stretching upward and outward in an amazing variety of patterns against a somber sky. There are sounds and sights that belong especially to this season of the year.

This is a particularly good time to study trees and shrubs. Donald Peattie, in his delightful book, ***The Natural History of Trees of Eastern and Central North America*** has this to say about tree study: "The first reward of tree study - but one that lasts to the end of your days - is that as you walk abroad, follow a rushing stream, climb a hill, or sit on a rock to admire the view, the trees stand forth proclaiming their names to you. Though at first you may fix their identity with more or less conscious effort, the easy-to-know species soon become like the faces of your friends, known without thought and bringing each a host of associations." This quotation can be applied to both tree and shrub study at any season of the year, but at no time better than fall.

I consider myself fortunate in living within a mile and a half of a nature center that gives me a chance to enjoy tree and shrub study the year around. There is a hill to climb, a view, and a stream that runs through a sizeable portion of the nature center property. A variety of different wildlife habitats are easily available for study . . . an old field returning to woodland, a young forest of gray birch, a wetland area and a hillside of oaks and hickories. All of these habitats are along a loop trail that is less than two miles. Let's explore this trail together.

Perhaps the most spectacular discovery that we will make on our walk is the late-flowering witch hazel with its ribbon-like blossoms. After other trees and shrubs have long since flowered and produced fruit, witch hazel produces a bloom. It is sometimes called "winterbloom" because, on some occasions, it waits until snow is on the ground before it flowers. The explosive fruit does not appear until a year later.

We follow the trail through a heavy stand of mountain laurel with its waxy evergreen leaves. This brings us to the foot of the hillside. Here, we can pause on our walk to take a look at tree silhouettes, winter buds and the barks of trees. Notice the muscular bark of the American hornbeam, the ski-trail characteristics of the bark of the red oak, the tattered bark of the shagbark hickory and the alligator-patterned bark of the flowering dogwood. The branches of the flowering dogwood are unusually picturesque as the dome-shaped winter flower buds stand out at the ends of the branches curving upward against the sky.

On the hillside, the dominant trees are the oaks and hickories with the American hornbeam and the flowering dogwood playing a supporting role as understory trees. The prickly evergreen, the common juniper, furnishes a sharp contrast to the bare branches of the broadleaf trees. Its blue berries are aromatic and used for flavoring gin.

The path up the hillside is beginning to accumulate more than its share of leaves. Hidden beneath them is an abundance of acorns that feel like ball bearings underfoot as we make our slippery ascent.

At the top of Pigeon Hill, as it is called, it is time to get our breath. How good it is to be alive and surrounded by so much beauty. From this vantage point there is a good view. Many of the oaks still have their leaves and one oak tree has a squirrel nest near the top. There is the zee-zee-zee of a golden crowned kinglet, a tiny bird often found in the company of black-capped chickadees and tufted titmice, and the occasional plop of another acorn as it hits the ground.

Searching for a handful of acorns, we can see that there are a great variety of shapes and sizes. The red oak acorn is larger than the other species and has a shallow cup. Its taste is bitter because it is full of tannin. Two years are required for the red oak acorn to develop. Some species, like those of the white oak, mature in one year. The black oak acorn is smaller than that of the red oak, and the cup is shingled, covering more of the acorn. The third species that grows on the hillside is the scarlet oak, these are less common here. It has an acorn that often has concentric rings around the top of the nut. This is best seen with a hand lens and is not always present.

The trail leading back to the Center offers a good opportunity to study several species of birch by looking at their bark. The silver-to-yellow bark of the yellow birch has a tendency to curl. It is sometimes called curly birch. Black birch has bark that is dark gray to black and very tight, until it starts to age. At this time it begins to reveal many cracks. Both black and yellow birch have twigs that smell like wintergreen when they are broken. Black birch was once a major source for oil of wintergreen but now this is produced synthetically. Gray birch has tight grayish-to-white bark with conspicuous black, chevron-like marks below each branch. This feature is missing in white birch, which is less common at the Center. It is a more northern species that has white bark, and it is easier to peel than gray birch.

There are some interesting shrubs that grow along this portion of the trail, but their identification in late fall and winter is often difficult without close observation of its winter buds; however, if a few leaves remain or fruit are present, identification can often be established. Arrowwood, a viburnum, can be identified by a few isolated bronze or red serrated leaves and by its round or oval bluish fruit with deeply grooved seeds. Winterberry, a common deciduous holly, can usually be identified by its spectacular red fruit that can be seen more dramatically after the leaves are shed. The fruit, attached directly on the stem, are usually in clusters, but some may occur singly. Two other common shrubs that have red berries are spicebush and Japanese barberry. Spicebush has an aromatic smell to the fruit and Japanese barberry has unbranched spines and all its berries occur singly.

The trail follows along the stream to the wooden bridge that leads back to the Center. As we stand here and listen to the water trickling under the bridge, here and there in the stream we see fallen leaves forming colorful little dams.

Much of the area around us is wetland. It has its share of swamp red maple, with its pale gray bark, and spongy sphagnum moss, now bleached nearly white. In late spring and summer, this area is a virtual jungle of cinnamon fern. They are now different tones of lacy brown, hugging the ground in twisting patterns.

As we approach the Center, we pass by a beautiful stand of Autumn olive with its gray-green leaves still present but its branches loaded down with hundreds of reddish berry-like fruit. These shrubs, or small trees, provide a feast for many species of birds, as we can see and hear, but in some places, they can be a real nuisance, crowding out other species.

What makes walking in the woods a unique experience is the variety of habitats usually found in close proximity to each other, and this will vary from woods to woods. Habitats often differ in their soil conditions, amount of moisture, degree of slope, amount of sunlight and associated plants. We have only experienced a few of the habitats found in these particular woods. Because every woods is different, this adds to the excitement when entering a new forest environment.

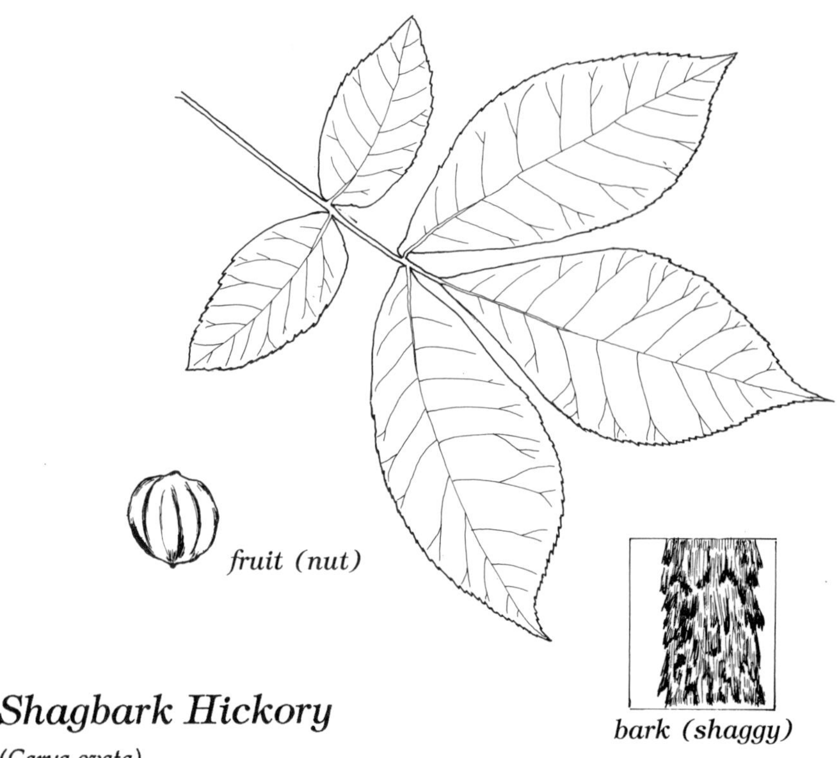

Shagbark Hickory

(Carya ovata)
Walnut family

Size: 60-90′

Few trees have bark that catches the eye quite like that of Shagbark hickory, a tree often found growing on dry upland slopes in association with oaks. The long, loose, curled, smoke-colored strips of the mature shagbark can leave a lasting impression. Young trees have smooth bark, the shagginess usually appearing when the tree is about 40 years old.

The winter buds are another noticeably feature of this species, particularly when they swell in the spring. Out of these buds come the folded, compound leaves and its flowers in the form of catkins. Both male and female catkins appear on the same tree, sometimes as early as April.

The edible nut has a heavy husk that splits nearly to the base when it ripens. Hickories are sometimes confused with walnuts but the leathery fruit of walnut never splits. Another difference is that the young twigs of hickory have solid pith, while walnuts have chambered pith.

The word hickory is an Indian word derived from "hickory milk," a substance obtained from the nut and once used in cooking.

Shagbark hickory wood can stand stress, a feature useful in making tool handles, baseball bats and skis. A small bird, the brown creeper, sometimes builds a nest under the loose bark, while many birds feed on the nuts.

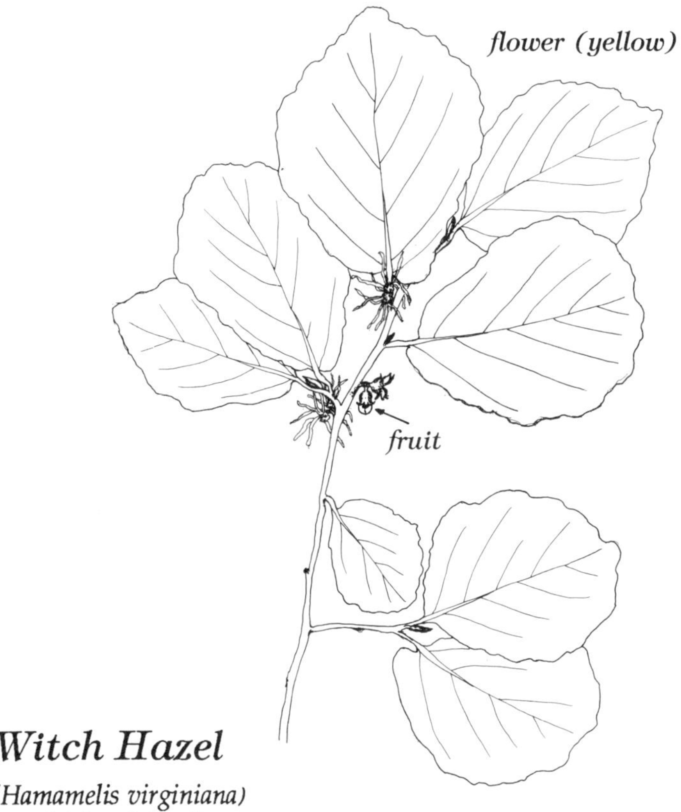

Witch Hazel

(Hamamelis virginiana)
Witch Hazel family

Size: 10-25′

Witch hazel is a shrub that likes to be different. When the leaves of most broadleaved trees and shrubs have begun to turn color and are dropping to the ground, it begins to flower. This occurs from October to December.

The bright yellow blossoms have four ribbon-like and twisted petals. Its fruit is in the form of a capsule taking one year to develop after the flower is formed. This capsule is able to discharge its small black, football-shaped seeds in an explosive manner, to a distance of up to 45 feet.

Witch hazel can grow in a wide variety of climatic conditions but is often found growing along the edges of woods and fields.

It is well known for the witch hazel lotion made from the twigs and bark and is still used for the treatment of bruises, bites and muscular stiffness. Historically, the forked branches of witch hazel were used as divining rods, believed to be a way of locating underground water.

A number of birds and mammals make use of the seeds, buds and twigs as part of their food supply. It is the preferred food for ruffed grouse, bluejays and catbirds.

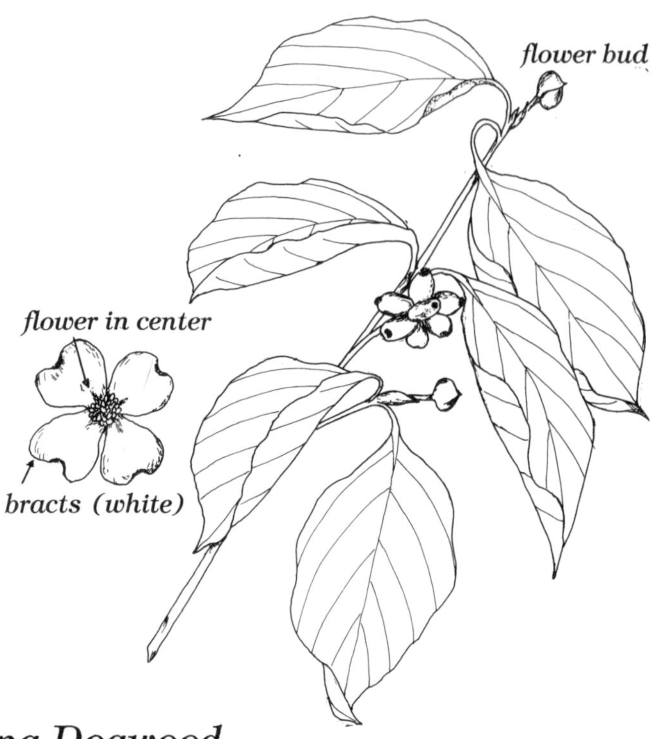

Flowering Dogwood

(Cornus florida)
Dogwood family Size: 10-40′

Flowering dogwood with its showy flowers and alligator-like bark is a common understory tree. It grows in a variety of soils, but it prefers rich, moist sites.

The flowers from this tree emerge from dome-shaped winter buds between March and July, the leaves developing from another type of winter bud. Four white, or sometimes pink, petal-like bracts surround the numerous tiny flowers or florets that make up the center. The colorful bracts attract insects to the nectar in the florets.

Fall brings a change in the appearance of the tree with its cluster of oval red fruit and colorful leaves varying from rose to scarlet to purple.

The genus name *Cornus* is derived from the Latin word for "horn." This refers to the hardness of the wood which is tougher than hickory. The common name was once "dagwood" because of its early use in making daggers and skewers.

The ability of this hard wood to resist shock makes it desirable for heads of golf clubs, chisel heads, small pulleys and shuttles for the textile industry. The fruit, although poisonous to man is eaten by over 100 species of birds, while white-tailed deer use the twigs and leaves for browse.

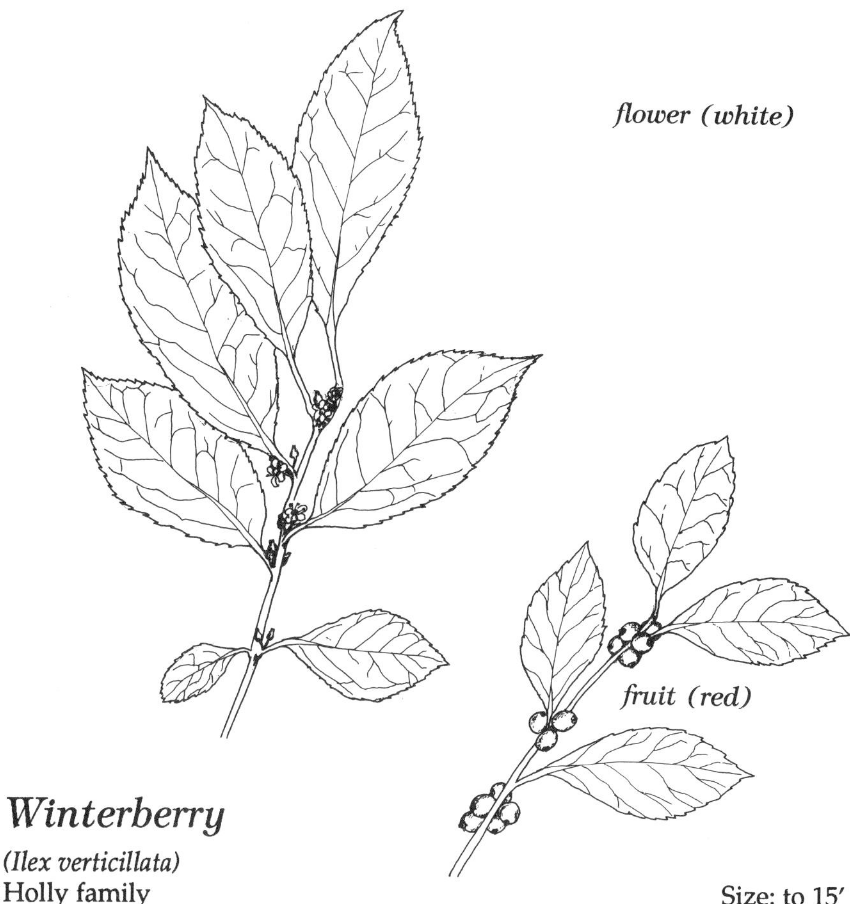

Winterberry

(Ilex verticillata)
Holly family

Size: to 15′

Winterberry is one of our native hollies and is widespread throughout the Northeast. It tends to go unnoticed during the spring and summer because the small greenish-white flowers are very inconspicuous; however, in the fall its bright red berries stand out in the autumn woods. They provide a dash of color to the forest at a time when the autumn foliage is past its peak, and they persist into the winter.

Winterberry is often found growing in wet woods or near swamps, ponds and streams. It is sometimes called black alder because its leaves turn black in the fall and it often grows in a similar environment to that of alders. Unlike many hollies that are evergreen, this holly sheds its leaves in the fall.

When we think of holly, it is usually of the American holly *(Ilex opaca)* closely resembling the European holly *(Ilex aquifolium)*. Both of these species are commonly used for Christmas decorations; however, winterberry is also valued for this same purpose. During the summer months, the leaves of winterberry can be made into a drink if they are boiled and then cooled. The berries are considered poisonous to man, but are the preferred food for the mockingbird, catbird, brown thrasher and the hermit thrush.

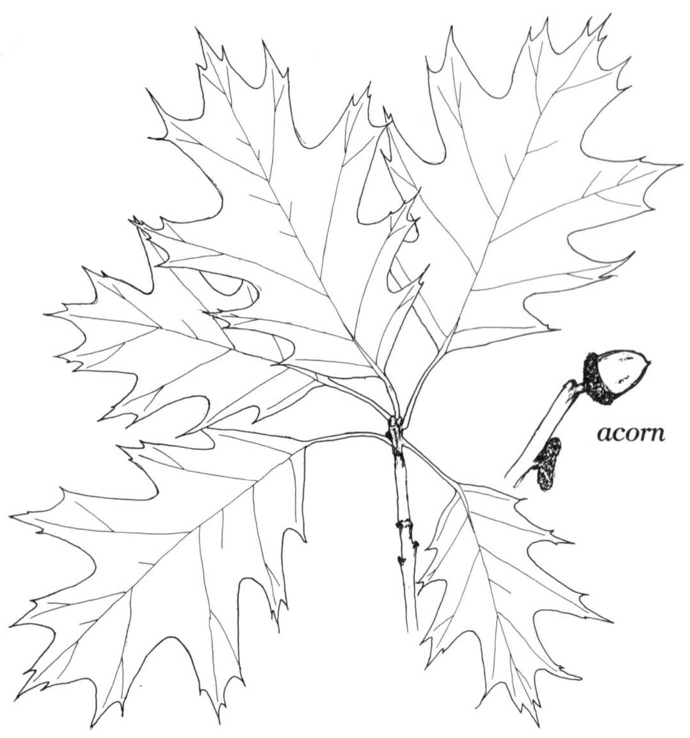

Northern Red Oak

(Quercus rubra)
Beech family — Size: 70-80′

Oaks are usually divided into two groups, those with leaves that have bristle tips and those without these tips. Red oak has the characteristics of the first group, while the rounded leaves of white oak are representative of the latter.

Red oak commonly grows on hillsides at various degrees of slope. The flowers are catkins, forming on separate parts of the same tree and developing about the same time as the leaves. The leaves are pink when they are developing in the spring and later become a dull green. Autumn finds them a brown to reddish-orange color.

Red oak bark is dark brown and has shallow, vertical furrows resembling ski-trails. The fruit is an acorn taking two years to develop. The top portion is quite broad and is set into a shallow saucer.

The reddish wood, the reason for its common name, is valued as a source of lumber for flooring, fence posts, veneer, pulp and fuel. Red oak is a popular ornamental for streets and lawns and its acorns and twigs have considerable value for wildlife.

Chapter 14

Plants of the Autumn Woods

American Hornbeam - *Carpinus caroliniana*
Autumn Olive - *Elaeagnus umbellata*
Black Birch - *Betula lenta*
Black Oak - *Quercus velutina*
Common Juniper - *Juniperus communis*
Flowering Dogwood - *Cornus florida*
Gray Birch - *Betula populifolia*
Mountain Laurel - *Kalmia latifolia*
Northern Arrowwood - *Viburnum recognitum*
Northern Red Oak - *Quercus rubra*
Scarlet Oak - *Quercus coccinea*
Shagbark Hickory - *Carya ovata*
White Oak - *Quercus alba*
Winterberry - *Ilex verticillata*
Witch Hazel - *Hamamelis virginiana*
Yellow Birch - *Betula alleghaniensis*

15

A Walk in the Winter Woods

I ONCE OVERHEARD a person comment, "What in the world can anyone see in the woods in the winter?" The answer is that you can see a great deal.

One of the satisfying aspects of winter botany is that it is an excellent chance to learn more about plant adaptations in winter. Our walk today will take us through several different habitats where we can observe a number of plants in winter and study their adaptations to a cold environment.

The path that we take into the winter woods leads through an old field that is starting to return to the forest conditions. Old field, or gray birch, red cedar, black cherry and silky dogwood, all light-loving plants, are indicators of this process.

As we approach the edge of the woods a clump of staghorn sumac with its red fruit still persisting during the winter months, attracts our eye. Here, this plant is part of the ecotone, a term used to indicate an area where two different environments meet. In this case, it is the forest and the old field. An ecotone may produce what is known as the "edge effect," meaning that often more animal life is found here than in either the field or the deeper woods. These animals have the best of two worlds to satisfy their needs for food and shelter.

A short distance into the woods, the vegetation changes rather dramatically. The sun-loving gray birches are still present, but they are in various stages of disarray . . . the place looks like a disaster area with the birches bent over every which way and many of them covered with birch conks, or shelf fungi, hastening the process of decay. A jumble of catbrier is intertwined with the birch. Red maple is pushing its way skyward to seek its place in the sun at the expense of the birch. The shade-tolerant maple, has gradually become successful as the predominant species, where it competes with the birches for sunlight. The birches, unable to stand the shade, are gradually losing their struggle for survival.

On the opposite side of the path, picturesque pitch pine with gnarled and drooping branches are present in some abundance. They too are sun-loving plants, but they have not yet run into any competition, this would then, result in their fate being the same as that of the birches.

A bit further along the path, the trail divides. We pause for a minute to inspect the remains of a clump of Indian pipe in its winter guise. The stems, now erect, are thin, black and wire-like with tiny urn-shaped tops containing the seeds. On the other side of the path, near a fork, Prince's pine or pipsissewa, with its woody stem and shiny, waxy green leaves, pokes up through the snow. These features, along with its small size growing near the ground, are all adaptations of evergreen plants to cope with an inability to obtain new supplies of water when the ground is frozen, and help cut down on evaporation by protecting the plant from the drying effects of wind. In order to prevent freezing, the amount of water in evergreen leaves in winter becomes somewhat reduced, and the sugar content is increased. This serves as a natural anti-freeze because it lowers the freezing point of the remaining water. This gives the plants protection from the killing frost.

Let us walk down the trail to renew old acquaintances with a white, or canoe birch. Rub your hand up and down the tight snow-white bark. Observe the white powder that easily coats your hand. This is a simple test we can use to tell this species from the gray birch it resembles. White birch also lacks the black chevrons that are typically below the branches of the gray birch.

Gazing down the trail leading to the brook, we see yellow birch with its curly bark and an occasional American hornbeam with its muscle-like trunk, characteristic of a wetter site.

Backtracking to the fork in the trail and proceeding along the path to the left, dry snow crackles under our feet and we move into a different part of the forest. Here, the trees are rather sparse with oaks and hickories predominant. These are more mature forest trees usually preferring hillsides and drier soil.

A fallen oak log near the edge of the trail reveals an unusual looking shelf fungus. This is the gill-maze Daedalea whose underside has an interesting maze-like pattern. This is where the spores lie hidden. These fungi, like the birch conks mentioned earlier, are considered decomposers. They occupy an important role, or niche, in the forest ecosystem. They eventually reduce wood to nutrient-rich soil that will nourish new green plants, the producers, which in turn provides food for animals, the consumers. Thus the cycle is complete and is constantly being repeated if the natural conditions remain the same.

Winter is a splendid time to learn more about ferns and the closely related clubmosses. Several species of our native ferns are evergreen, and all our clubmosses retain their greenery all year long.

Christmas fern, a couple of species of wood fern and rock fern are the common evergreen ferns that grow along the trail. Of these, Christmas fern grows in the greatest abundance. It sometimes grows in colonies, but more often singly or in two's or three's. The common name comes from its early use for Christmas decorations and perhaps because each leaflet, or pinna, resembles the foot of Christmas stocking.

Clubmosses are small evergreen plants that have either an upright or trailing growth habit. Most of them have straight stems with club-like cones. This, and their somewhat mossy appearance is the reason they are referred to incorrectly as clubmosses. They are not related to true mosses. Some of them look like tiny pine trees, others have leaves that appear pressed, like small cedars. They usually grow in colonies.

Both ferns and clubmosses produce spores, but their most rapid and certain way of reproducing is by the rootstalk, a kind of stem that grows horizontally either above or below the ground. The older growth withers and dies but growth is more rapid than death. The result is that the size of the colonies tends to increase.

Ferns and clubmosses have an interesting fossil history that extends back into geologic times for more than two hundred million years. They were much larger then, growing to tree size, and they made up a significant part of the vast forest jungle. Despite their ancient origin, ferns and clubmosses do not have an unusually long life span, but there are clubmoss clones in the forest that are many decades old. The distinction of being the oldest living thing on earth goes to the creosote bush. Old creosote bushes are also clones with dead centers.

Tree clubmoss, ground-cedar, shining clubmoss and wolf's-claws are among some of the commonly found clubmosses growing in colonies here in the winter woods.

The trail winds downhill and leads nearer the brook. Here we stop again to look at the bare stems of the black-berried elderberry. Run your hand up and down the stem to feel the warty lenticels that are characteristic of this plant. Lenticels are pores that the plant uses for gaseous exchange in the winter when the leaves are no longer present to perform this function.

Now that the leaves are gone, winter buds become more evident. With the help of a hand lens we can distinguish the features that make each bud unique although at first glance they may look much alike. One bud that is intriguing is that of the tulip tree, having terminal buds with only two scales, giving it the appearance of a duck's bill. A thin, zig zag branch of a nearby American elm attracts our attention. Its buds are pointed and sit at an abrupt angle to the stem, looking much like dunce caps. These buds contain miniature leaves formed sometimes during the summer and will emerge next spring.

We are just about ready to start back home when we notice a type of jelly fungus called "witches butter" growing on a fallen log. Notice the yellow gelatinous, lobed and brain-like mass. It is more a part of the winter than the summer woods because of its particular moisture and temperature for growing and producing fruiting spores. Most fungi fruit at other times of the year. The spores of this unusual fungus are borne on the upper surface of the lobes rather than on the undersurface, as in the shelf-like Daedalea mentioned earlier.

Darkness is beginning to move gradually over the woods as we begin to walk briskly back to our starting point, pausing only for a moment to take a quick look at a patch of rattlesnake plantain, one of our native orchids we overlooked. This plant has leaves with a striking checkered white and green pattern.

Glancing at the nearby brook, we notice that there are breaks in the ice and the water bubbles out for a few moments only to disappear again under the ice. The sound of the water as it bubbles up between the cracks in the ice breaks the pervading silence of the winter woods, and I ponder about how much we have seen and heard on our walk in this woodland habitat.

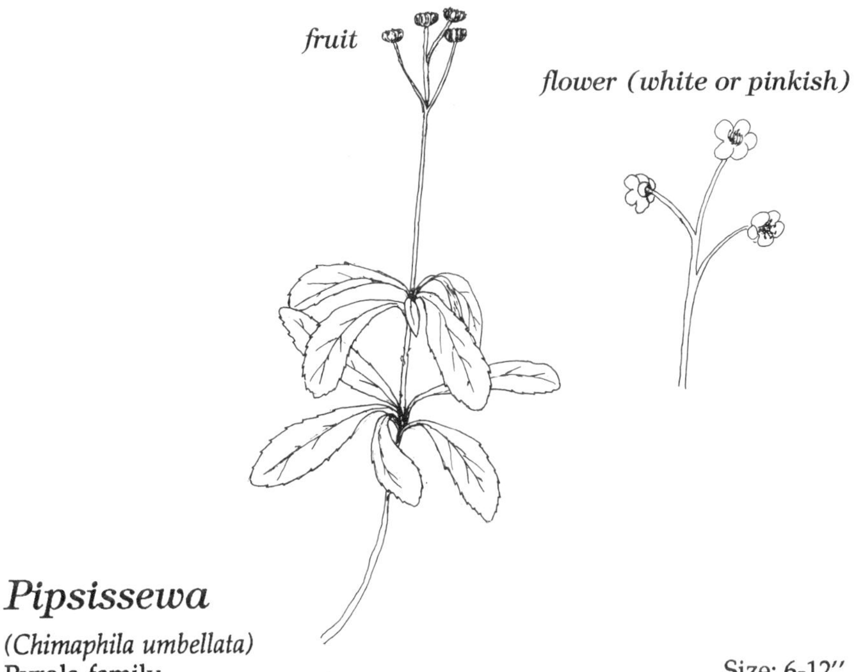

Pipsissewa

(Chimaphila umbellata)
Pyrola family

Size: 6-12″

Pipsissewa, or Prince's pine, is a delightful plant to discover in the winter woods. It is considered a semi-shrub and is especially attractive with its whorl of shiny, leathery, green leaves. The generic name of this species is a combination of two Greek words meaning "winter loving," referring to the green leaves remaining green throughout the winter months.

Pipsissewa is most commonly found growing in shade in dry woods under cone-bearing trees like pine, spruce and hemlock. The erect stem, growing up to 12 inches in height, develops from an underground, creeping stem (rhizome). New shoots develop at intervals from the rhizome to produce a colony of plants.

During July and August, a beautiful wax-like flower appears. The white to pink flowers form terminal clusters, umbrella-shaped. The species name, *umbellata* describes this feature. Each individual flower has five widely separated petals, ten hairy stamens and a single, short green style with a disk-like stigma. Striped or spotted wintergreen *(Chimaphila maculata)* is a common, closely related species having leaves striped with white.

Pipsissewa is an Indian name. In Cree language it means "juice breaks stones of bladder into small pieces." This is a reference to its use for treatment of gallstones, kidney stones and other urinary ailments.

The American Indians also used this plant as a treatment for rheumatism and skin disorders.

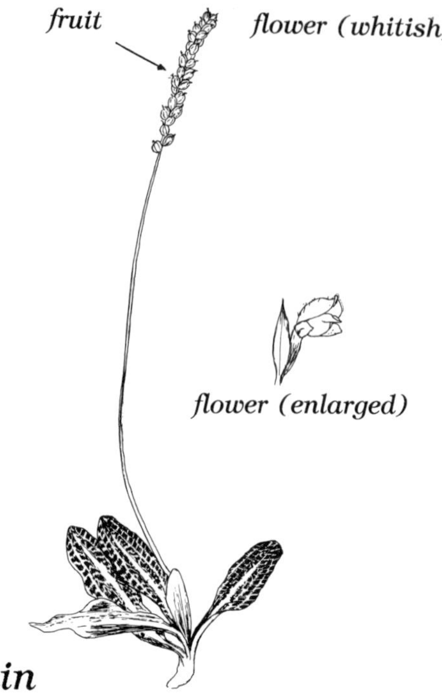

Rattlesnake Plantain

(Goodyera pubescens)
Orchid family

Size: to 18″

Rattlesnake plantain is a handsome plant even when it is discovered during the winter. The leaves are its most attractive feature, being arranged in a basal rosette and having a beautiful checkered pattern of white veins against a dark green background. The leaves may be covered with snow, but the flower stalk with the remains of the flowers and seeds may be visible throughout the winter months.

Rattlesnake plantain is not a plantain, but one of our native American orchids. It is a plant of dry oak or pine woods. The genus name is in honor of John Goodyear, an English botanist of the seventeenth century. *Pubescens,* its species name refers to the downy nature of the stem covered with glandular hairs.

Its white to greenish flowers take the shape of a dense spike-like cylindrical cluster that blossoms from July to September. The colored upper sepal joins with two petals to form a hood covering the third petal acting as a sac-like lip.

Indians thought its leaves resembled a snake skin and was therefore useful for snake bites. The root was chewed and part of it was swallowed, while some was applied to the wound.

A solution made from the leaves was once used in treating skin ailments.

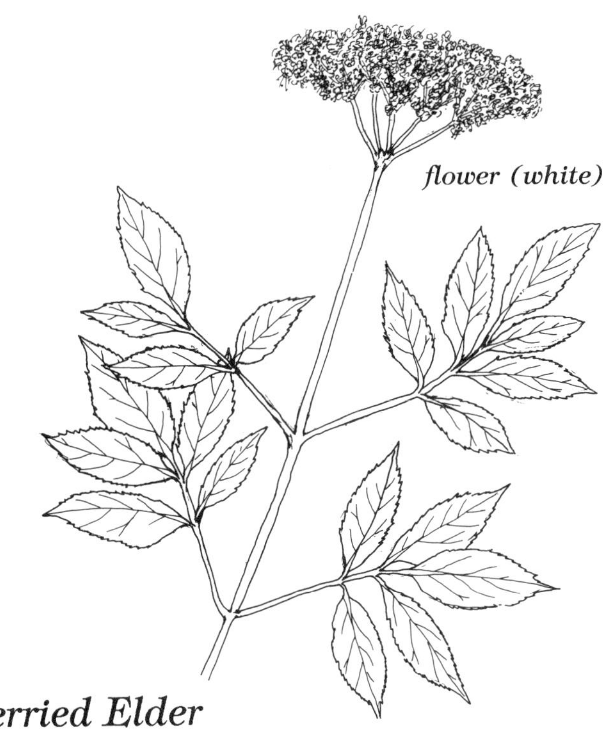

Black-berried Elder

(Sambucus canadensis)
Honeysuckle family

Size: 3-13′

Black-berried elder graces the landscape in a variety of open habitats such as roadsides, forest edges, along streams, and it often forms thickets.

The flower of this species is quite attractive, appearing from June to July in white flat-topped clusters. Each flower of the cluster is only about 1/4 inch across and consists of five small petals and five distinct stamens. Its fruit is purplish-black and berry-like.

The bark of this tree is covered with warty lenticels, these are raised pores that help the plant in gaseous exchange during the winter months when the leaves are not present. This species has white pith in the center of the stem, a characteristic differing from the red-berried elder *(Sambucus pubescens)* which has brown pith.

Its leaves are compound with usually about seven leaflets, but the number may vary. All parts of the plant have at least some hydrocyanic acid present. This is destroyed by cooking.

The flower clusters, called elderblow, can be dipped in batter to make fritters, while the fruit is used in making wine, jams and fillings for pies. Whistles, popguns and spikes for tapping maple trees can be made out of the easily hollowed-out stem.

At least 47 species of birds feed on the fruit.

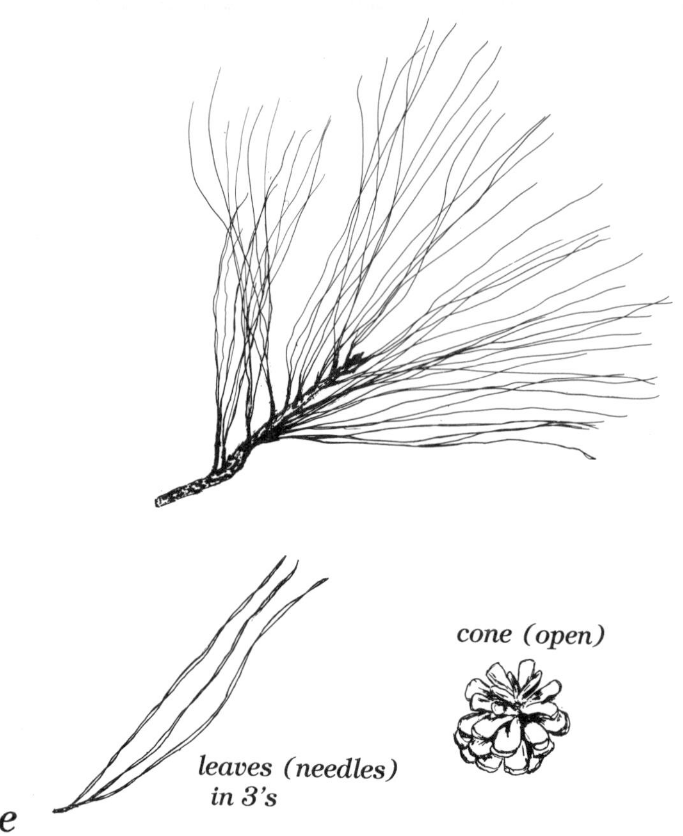

Pitch Pine

(Pinus rigida)
Pine family

Size:40-60′

Pitch pine's growth habits in the Northeast are somewhat picturesque, its branches often creating a gnarled and drooping effect. The pitch pine cones also enhance its appearance as they persist on the branches for several years after they have dropped their seeds.

This species of pine is often found growing in open areas on dry, sandy or sterile soil. Because it is a sun-loving species, it does not grow well or reproduce under forest conditions. It is one of the few evergreens capable of sprouting after a fire. This accounts for the abundance of pitch pine in the New Jersey pine-barrens.

Pitch pine has three needles in each bundle and the needles are often twisted. White pine differs in having five needles.

The wood of pitch pine makes good fuel and is a fine source of charcoal. Its lumber is used for coarse construction. The early colonists used it for shipbuilding. During the nineteenth century, it furnished the boilers of woodburning locomotives with large amounts of firewood.

Hairy woodpeckers, black-capped chickadees and white-breasted nuthatches use pitch pine for food, cover and for their nesting sites. White-tailed deer use the young sprouts and seedlings as browse.

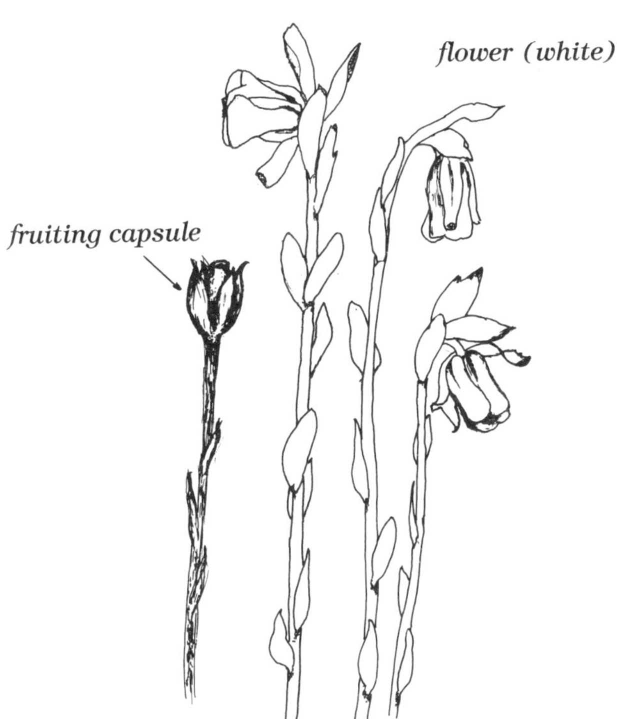

Indian Pipe

(Monotropa uniflora)
Wintergreen family

Size: 3-9″

Indian pipe is also called corpse plant and ghost plant. Both names allude to the eerie appearance of this unusual plant. Its most noticeable feature is the lack of cholorphyll. However, it is considered a flowering plant belonging to the wintergreen family.

The mystery of how it can function as a flowering plant can be solved by an inspection of its root system. This consists of a spiny ball with spines covered with a fungus growth *(mycorrhizae)*. This type of soil fungus is able to digest the decaying vegetative matter surrounding it and deliver it to the cells of the Indian pipe.

Indian pipe is found growing in shady areas, more often under pines than broadleaved trees. It generally appears in clusters, arising from a compact root system. It is drooping and pipe-like in appearance as it grows upward and blossoms. This occurs between June and September. The single waxy flower is usually white but may be pink. As the urn-shaped fruit develops, the flower becomes erect and turns black. With the approach of winter, the plant loses its scale-like leaves and becomes wiry looking. In this form, it persists throughout this season.

The Indians and early herbalists used the juice from the Indian pipe as an eye-lotion and the dry herbs for treating nerve ailments.

Chapter 15

Plants of the Winter Woods

American Elm - *Ulmus americana*
American Hornbeam - *Carpinus caroliniana*
Birch Conk - *Piptoporus betulinus*
Black-berried Elder - *Sambucus canadensis*
Black Cherry - *Prunus serotina*
Catbrier - *Smilax rotundifolia*
Christmas Fern - *Polystichum acrostichoides*
Gill-maze Daedalea - *Daedalea quercina*
Gray Birch - *Betula populifolia*
Ground-cedar - *Lycopodium flabelliforme*
Indian Pipe - *Monotropa uniflora*
Pipsissewa - *Chimaphila umbellata*
Pitch Pine - *Pinus rigida*
Rattlesnake Plantain - *Goodyera pubescens*
Red Cedar - *Juniperus virginiana*
Red Maple - *Acer rubrum*
Rock Fern - *Polypodium virginianum*
Shining Clubmoss - *Lycopodium lucidulum*
Silky Dogwood - *Cornus amomum*
Staghorn Sumac - *Rhus typhina*
Tree Clubmoss - *Lycopodium obscurum*
Tulip Tree - *Liriodendron tulipifera*
White Birch - *Betula papyrifera*
Witches Butter - *Tremella mesenterica*
Wolf's-Claws - *Lycopodium clavatum*
Wood Fern - *Dryopteri* sp.
Yellow Birch - *Betula alleghaniensis*

Parts of a Simple Flower

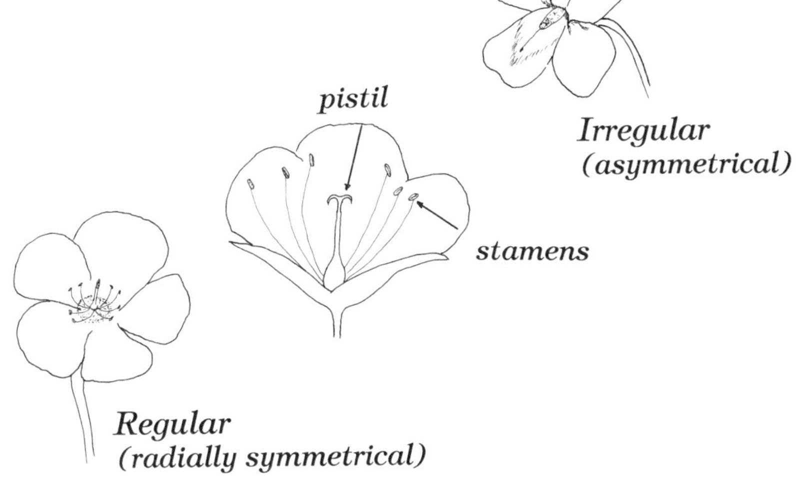

Parts of a Composite Flower

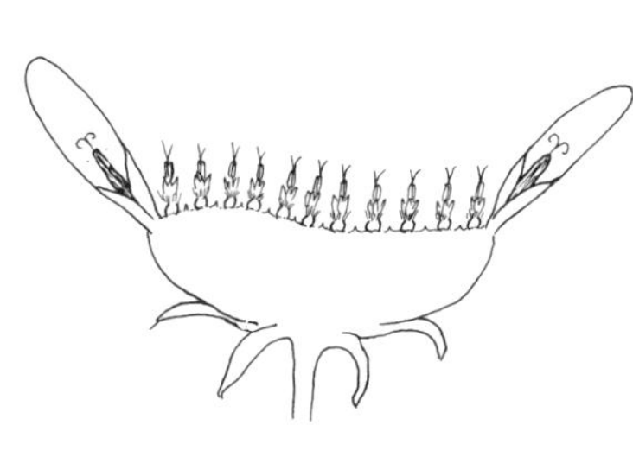

Ray flower

Glossary

ACHENE - a dry thin-walled fruit containing a single seed.

ANTHER - the part of the stamen that produces pollen and is supported by a thread-like filament.

BLOOM - a white powdery covering, often wax-like.

BRACTS - small modified leaves that are often located below a flower or cluster of flowers.

CALYX - a collective term for the sepals, usually green, found below the petals of a flower.

CATKINS - a dry, scaly, unisexual spike that is usually flexible - common in willows, oaks, birches and poplars.

CAPSULE - a dry fruit that splits into two or more parts when it ripens.

CAUSEWAY - a road constructed over a wet or marshy area.

CLEISTOGAMOUS FLOWERS - small self-fertilized flowers that do not open, and produce seeds.

CLONE - a group of cells that have the same genetic makeup, descending from the same common ancestor.

COMMUNITY - the plants and animals of a given area that act as an interrelated group; e.g. all the plants and animals of a beech-maple forest.

COMPOSITE FLOWER - a flower head composed of many small flowers or florets, surrounded by bracts.

COMPOUND LEAF - a leaf consisting of two or more leaflets.

COROLLA - the collective name for the petals.

CULTIVARS - varieties of plants that have arisen under cultivation.

DISK FLOWER - one of the small tubular flowers found in the center of the head of a composite.

DIMORPHIC - referring to a flower that may appear in two forms.

DIURETIC - a substance used to aid the flow of urine.

DRUPE - a stone fruit that is fleshy and one-seeded.

ECOTONE - an area where two communities come together and intermingle.

EDGE EFFECT - the rise in population of animal life where two communities meet.

EMERGENT - a rooted water plant that extends above the water surface.

HEAD - a spike that is short and thick, consisting of a dense cluster of stalkless or nearly stalkless flowers.

INFLORESCENCE - the manner in which the flower cluster is arranged.

INROLLED - a term used to describe the margins of leaves when they are curled inward - an adaptation to conserve on water.

IRREGULAR FLOWER - asymmetrical - with some petals differing from others in size and shape.

LENTICELS - small corky spots that allow for gaseous exchange in young trees and shrubs.

MICRO-CLIMATE - the weather conditions in the first six feet above the ground.

MUSKEG - when a lake becomes completely filled in with sphagnum moss and associated woody plants.

MYCORRHIZAE - a beneficial relationship between certain soil fungi and plant roots.

NATURALIZED - a term used to characterize species of plants that have become permanently established in a new area.

NECTARIES - the nectar producing part of a flower.

NITRATES - a form of nitrogen most commonly used by green plants.

OVARY - the lower part of the pistil that contains the ovules, unripe seeds.

PAPILIONACEOUS - flower that appears butterfly-like, as in peas.

PAPPUS - hairs, bristles or scales that are present on the summit of the "seed-like" achenes of the composite family.

PISTIL - female reproductive organ consisting of stigma, style and ovary.

RACEME - an elongated inflorescence with individual stalked flowers.

RAY FLOWERS - the outer petal-like flowers of some composites.

RECURVED - curved backwards.

REGULAR FLOWER - radially symmetrical - with all petals like each other in size and shape.

RHIZOME - an underground stem or rootstalk, usually horizontal.

ROSETTE - a cluster of leaves that radiate outward, usually basal.

SAPONINS - chemicals that are toxic (glucosides).

SEPALS - individual parts of the outer whorl of a flower (calyx), usually green.

SPADIX - a fleshy spike surrounded by a large leaf-like bract - the spathe.

SPRING EPHEMERALS - plants that bear spring flowers for a short time and then the leaves along with the flower disappear, leaving only the underground storage part of the plant.

SPIKE - an elongated inflorescence with individual flowers that are stemless or nearly so.

SPUR - a hollow tube or sac-like elongations of some portion of a flower, usually containing nectar.

STAMENS - the male organs of a flower that produce pollen.

STIGMA - the upper part of the pistil where pollen is received.

STIPULES - leaf-like parts often located at the base of a petiole or leaf.

STOLONS - a horizontal stem such as a runner that tends to produce new stems at the tip.

STYLE - a part of the pistil located between the stigma and the ovary.

SUBSTRATE - the underlying rock, soil or water medium.

TENDRILS - modified leaves that clasp or twine around an object, used for support or climbing.

THALLUS - a plant body that has not become differentiated into stem or leaf.

UMBEL - a flower with all its florets coming from the same point, like the umbrella.

VEGETATION DEVELOPMENT - a continuous series of natural changes in a wildlife community following a disturbance of some kind, such as fire, wind or man's intervention, probably leading to a mature vegetation type if there is no further interruption.

VEGETATIVE PARTS - non-reproductive parts of a plant.

VEGETATIVE REPRODUCTION - reproduction without sexual reproduction, such as with bulbs, runners and tubers.

WHORLED - three or more leaves arising from the same point on a plant.

Index

Index

Index

Index

Index

Bibliography

Bailey, L.H., **Manual of Cultivated Plants,** New York, MacMillan Publishing Co., 1977

Barth, Friedrich G., **Insects and Flowers,** Princeton, N.J., Princeton University Press, 1985

Crockett, Lawrence J., **Wildly Successful Plants, A Handbook of North American Weeds,** New York, Collier Books, a division of MacMillan Co., 1977

Dalton, Patricia A. & Alejandro Novelo R., **Aquatic and Wetland Plants of the Arboretum,** Boston, Arnoldia, 1983. 43 (2): 7-44.

DeGraaf, Richard M & Witman, Gretchin M., **Trees, Shrubs and Vines for Attracting Birds,** Amherst, MA., University of Massachusetts Press, 1979

Dowden, Anne Ophelia T., **Look at a Flower,** New York, Thomas W. Crowell Co., 1963

Dwelley, Marilyn J., **The Trees & Shrubs of New England,** Lewiston, ME., Twin City Printery, 1980

Elias, Thomas S., **Complete Trees of North America,** New York, Van Nostrand Reinhold Co., 1980

Erichsen-Brown, Charlotte, **Use of Plants of the Last 500 Years,** Aurora, Ontario, Breezy Creek Press, 1979

Fernald, M.L., **Gray's Manual of Botany, Eighth Edition,** New York, D. Van Nostrand Company, 1950

Gibbons, Euell, **Stalking the Wild Asparagus,** New York, David McKay Company Inc., 1974

Gleason, Henry and Cronquist, Arthur, **Manual of Vascular Plants of Northeastern United States and Canada,** New York, D. Van Nostrand Company Inc., 1965

Godfrey, Michael A., **A Closer Look,** San Francisco, Sierra Club Books, 1975

Haughton, Claire Shaver, **Green Immigrants,** New York, Harcourt Brace Javanovich, 1978

Headstrom, Richard, **Nature in Miniature,** New York, Alfred A. Knopf, 1968

Headstrom, Richard, **Suburban Wildflowers,** Englewood, N.J. Prentice-Hall Inc., 1984

Johnson, Charles W., **Bogs of the Northeast, Hanover and London,** University Press of New England, 1985

Jorgensen, Neil, **A Sierra Club Naturalist's Guide to Southern New England,** San Francisco, Sierra Club Books, 1978

Kavasch, E. Barrie, **Guide to Northeastern Wild Edibles,** Blaine, Washington, Hancock House Publishing Inc., 1981

Klimas, John and Cunningham, James A., **Wildflowers of Eastern America,** New York, Alfred A. Knopf, Inc., 1974

Klots, Elsie B., **The New Field Book of Freshwater Life,** New York, G.P. Putnam's Sons, 1966

Lamoureux, W. John, **Aquatic Plants for Fish and Wildlife,** Technical Bulletin 1. Hamilton, Ontario, Royal Botanical Garden, 1970 (third edition)

Lees, David, **Great Bogs of Fire,** Pages 37-42, Harrowsmith, Feb/Mar 1983

Little, Elbert L., **The Audubon Society Field Guide to North American Trees - Eastern Region,** New York, Alfred A. Knopf, 1980

Martin, Alexander C., Zim, Herbert S., and Nelson, Arnold D., **American Wildlife and Plants,** New York, McGraw Hill Book Company Inc., 1951

Mathews, F. Schuyler, **Field Guide of American Trees and Shrubs,** New York, G.P. Putnam's Sons, 1915

Milne, Lorus and Margery, **Because of a Flower,** New York, Athenaum, 1975

Milne, Lorus and Margery, **Living Plants of the World,** New York, Random House, 1972

Minister of Supply and Services, **Weeds of Canada,** Ottawa, Canada, 1977

Niering, William, **Life of the Marsh,** New York, McGraw Hill Book Company, Inc., 1966

Niering, William and Olmstead, Nancy C., **The Audubon Society Field Guide to North American Wildflowers - Eastern Region,** New York, Alfred A. Knopf, 1979

Niering, William and Goodwin, Richard H., **Energy Conservation on the Home Grounds,** New London, CT., Connecticut Arboretum, 1975

Newcomb, Lawrence, **Newcomb's Wildflower Guide,** Boston, Little Brown and Company, 1977

Page, Nancy M. and Weaver, Richard E., **Wild Plants in the City, New York, Quadrangle-New York Times Book Co., 1975**

Palmer, E. Lawrence, Lawn Laboratories - Cornell Rural School Leaflet, Volume 42, Number 2, Fall 1948, Ithaca, New York, New York State College of Agriculture at Cornell University, 42(2): 1948

Peattie, Donald Culross, **A Natural History of Trees of Eastern and Central North American,** New York, Bonanza Books, 1963

Peterson, Lee, **A Field Guide to Edible Wild Plants of Eastern & Central North America,** Boston, Houghton Mifflin Co., 1978

Peterson, Roger Tory, and McKenny, Margaret, **A Field Guide to Wildflowers,** Boston, Houghton Mifflin Co., 1968

Petrides, George A., **A Field Guide to Trees and Shrubs,** Cambridge, MA., The Riverside Press, 1958

Pringle, James S., **The Common Solidago Species of Southern Ontario,** Hamilton, Ontario, Royal Botanical Gardens Technical Bulletin No. 3, 1968

Pringle, James S., **The Common Asters of Southern Ontario,** Hamilton, Ontario, Royal Botanical Gardens Technical Bulletin No. 2, 1981

Rickett, Harold William, **The New Field Book of American Wildflowers,** New York, G.P. Putnam's Sons, 1963

Rickett, Harold William, **Wild Flowers of the United States - The Northeastern States,** Vols. 1 & 2, New York, McGraw Hill Co., 1965

Roth, Charles E., **The Plant Observer's Guidebook,** Englewood Cliffs, NJ., Prentice-Hall Inc., 1984

Scott, Jane, **Botany in the Field,** Englewood Cliffs, NJ, Prentice-Hall Inc., 1984

Shosteck, Robert, **Flowers and Plants,** New York, Quadrangle-New York Times Book Co., 1974

Silverman, Maida, **A City Herbal,** New York, Alfred A. Knopf, 1977

Stokes, Donald W., **The Natural History of Wild Shrubs and Vines -Eastern and Central North America,** New York, Harper and Row, 1981

Stokes, Donald and Lillian, **Wildflowers,** Boston, Little Brown and Company, 1985

Symonds, George W.D., **The Shrub Identification Book,** New York, William Morrow & Company, 1963

Symonds, George W.D., **The Tree Identification Book,** New York, Quill, 1958

Teale, John & Mildred, **Life and Death of the Salt Marsh,** New York, Audubon/Ballantine Books, 1969

Viertel, Arthur T., **Trees, Shrubs & Vines,** New York, Syracuse University Press, 1970